Plant Tolerance to Individual and Concurrent Stresses

Muthappa Senthil-Kumar
Editor

Plant Tolerance to Individual and Concurrent Stresses

Springer

Editor
Muthappa Senthil-Kumar
National Institute of Plant Genome Research
New Delhi, India

ISBN 978-81-322-3704-4 ISBN 978-81-322-3706-8 (eBook)
DOI 10.1007/978-81-322-3706-8

Library of Congress Control Number: 2017930398

Printed on acid-free paper

This Springer imprint is published by Springer Nature
The registered company is Springer (India) Pvt. Ltd.
The registered company address is: 7th Floor, Vijaya Building, 17 Barakhamba Road, New Delhi 110 001, India

Preface

In nature, plants are exposed to one or more biotic and abiotic stresses either individually or in combination, which ultimately results in yield loss. During the life cycle, the same plant can face individually occurring one or more stresses. A large number of studies were undertaken to dissect the mechanisms imparting plant tolerance to multiple individual stresses. However, the concurrent stress tolerance has not been adequately studied owing to several complexities involved, including appropriate combined stress imposition method. In the recent past, several research groups around the world have started exploring the concurrent stress tolerance mechanisms under both biotic and abiotic stress combinations. This book compiles the information generated by these research groups along with their research progress and prospects, which would serve as a compendium of knowledge for researchers working on plant stress biology.

This book covers three major aspects under the proposed title. First, it introduces the existence of unique and shared responses in plants exposed to combined stress. Emphasis is given for understanding shared responses in comparison with multiple individual stresses. Second, the influence of abiotic stress on plant-pathogen interaction is elaborately covered. Third, comprehensive information about screening methods to identify genetic variation and the use of various tools to extrapolate information from individual stress studies to understand concurrent stress tolerance is elaborated. The chapterwise coverage of above said information is as follows.

Chapters 1 and 2 cover the overview of physiological and molecular mechanism involved in imparting both individual and combined stress tolerance. Importance is also given to the soil management and agronomic practices that will facilitate cultural management of crops under combined stress. Chapters 3 and 4 enumerates the impact of biotic stresses, namely, weed and pathogen on sequential and simultaneously occurring abiotic stresses including drought and temperature stress. Chapter 5 explains the approaches and avenues available for utilizing the understandings covered in the previous four chapters in terms of genomics-assisted breeding. Chapters 6 and 7 comprehend all previously described stress responses and set tone for specific stress tolerance mechanisms described in subsequent chapters. Chapter 8 focuses on the plant interaction with light and temperature, both as stimuli and stress. This chapter specifically covers the signaling responses and emphasizes the growth changes during combined stress. Hormonal cross talks under combined stress and the coordinated regulation of stress tolerance mechanisms are discussed in Chap. 9. Impact of several individual stresses on plants and strategies for crop improvement are covered in Chap. 10. The Chap. 11 covers the plant-water relations during various pathogen infections. It also enumerates the complexity of these responses in the presence of drought stress. Overall, these 11 chapters delivers scintillating information that not only provide comprehension of up-to-date research outcome in understanding stress interaction and combined stress tolerance, but also enumerate future direction of research. Overall they acts as suitable study material for both students and researchers working this area. This book also delivers prospects for driving future research for developing strategies for crop improvement under multiple stresses.

Eminent researchers from this newly emerging field have contributed to this book as outlined above. This book will be not only served as a one-stop reference point for researchers working in plant responses to both biotic and abiotic stresses but also will be an authority of recent information in this area. It is noteworthy to emphasize the fact that despite the plants grown under field condition exposed to combination of multiple stresses, a comprehensive collection of recent information in this area is lacking. This book will sufficiently address this deficit and act as a reference material for the research community.

I acknowledge all the reviewers who made scientific and technical comments on each chapter included in this book for their valuable time and input.

New Delhi, India Muthappa Senthil-Kumar

Contents

Contributors

Seth Bernard Abugho Department of Soil and Crop Sciences, Texas A&M University, College Station, TX, USA

Muthukumar Bagavathiannan Department of Soil and Crop Sciences, Texas A&M University, College Station, TX, USA

Juhi Bhattacharya National Institute of Plant Genome Research, New Delhi, India

Annvi Dhaka National Institute of Plant Genome Research, New Delhi, India

Urooj Fatima National Institute of Plant Genome Research, New Delhi, India

Raju Ghosh International Crops Research Institute for the Semi-Arid Tropics (ICRISAT), Patancheru, Hyderabad, Telangana, India

Prabhu Govindasamy Department of Soil and Crop Sciences, Texas A&M University, College Station, TX, USA

Aarti Gupta National Institute of Plant Genome Research, New Delhi, India

Sumandeep Juneja Centre for Plant Sciences, School of Basic and Applied Sciences, Central University of Punjab, Bathinda, Punjab, India

Sanjeev Kumar Centre for Plant Sciences, School of Basic and Applied Sciences, Central University of Punjab, Bathinda, Punjab, India

Anuradha Kumari Centre for Biochemistry and Microbial Sciences, Central University of Punjab, Bathinda, Punjab, India

Rui Liu Department of Soil and Crop Sciences, Texas A&M University, College Station, TX, USA

Mehanathan Muthamilarasan National Institute of Plant Genome Research, New Delhi, India

Gunasekhar Nachimuthu New South Wales Department of Primary Industries, Australian Cotton Research Institute, Narrabri, NSW, Australia

Karaba N. Nataraja Plant Molecular Biology Laboratory, Department of Crop Physiology, University of Agricultural Sciences, Bengaluru, India

Prachi Pandey National Institute of Plant Genome Research, New Delhi, India

M. S. Parvathi Plant Molecular Biology Laboratory, Department of Crop Physiology, University of Agricultural Sciences, Bengaluru, India

Manoj Prasad National Institute of Plant Genome Research, New Delhi, India

Wusirika Ramakrishna Centre for Biochemistry and Microbial Sciences, Central University of Punjab, Bathinda, Punjab, India

Aashish Ranjan National Institute of Plant Genome Research, New Delhi, India

Pranav Pankaj Sahu National Institute of Plant Genome Research, New Delhi, India

Muthappa Senthil-Kumar National Institute of Plant Genome Research, New Delhi, India

Mamta Sharma International Crops Research Institute for the Semi-Arid Tropics (ICRISAT), Patancheru, Hyderabad, Telangana, India

Priyanka Singh Centre for Plant Sciences, School of Basic and Applied Sciences, Central University of Punjab, Bathinda, Punjab, India

Roshan Kumar Singh National Institute of Plant Genome Research, New Delhi, India

Upendra Kumar Singh National Institute of Plant Genome Research, New Delhi, India

Vijay Singh Department of Soil and Crop Sciences, Texas A&M University, College Station, TX, USA

Ashley A. Webb New South Wales Department of Primary Industries, Tamworth Agricultural Institute, Calala, NSW, Australia

Renu Yadav Centre for Plant Sciences, School of Basic and Applied Sciences, Central University of Punjab, Bathinda, Punjab, India

About the Editor

Muthappa Senthil-Kumar is a scientist at the National Institute of Plant Genome Research, New Delhi, India. He received his B.Sc. in agriculture from Tamil Nadu Agricultural University, Coimbatore and M. Sc. and Ph.D. in crop physiology from the University of Agricultural Sciences, Bangalore, India. He was post-doctoral fellow at The Samuel Roberts Noble Foundation, Ardmore, Oklahoma, USA. He has published over 40 research articles and several review articles on understanding plant interaction with drought stress and pathogens. Currently, his research team is working to understand the interaction of drought and pathogen stress and their combined impact on plants.

1 Concurrent Stresses Are Perceived as New State of Stress by the Plants: Overview of Impact of Abiotic and Biotic Stress Combinations

Aarti Gupta and Muthappa Senthil-Kumar

Abstract

Crop plants under natural conditions often encounter abiotic and biotic stresses either individually or in combination, single or multiple times in their life cycle. During their concurrence, different stressors interact with each other over the plant interface leading to altered plant responses. Initial stressor can modulate plant physiology and thereby influences plant response towards another stressor. Consequent to the stress interaction, plants encountering concurrent stress show different responses in comparison to the plants exposed to the individual stresses. Additionally, plant defence responses are somewhat skewed towards one stressor during concurrent occurrence of stresses. Such different responses are the cognate 'net effect' of combined stress felt by the plant. The net effect exhibited by plants under combined stress is unique to each stress combination. Thus, in lieu of the combined stress responses, which are different from the individual stress responses, the combined stress has been proposed as a new state of stress. Plant responses towards this new state are not just dictated by either of the individual stresses alone but by more complex interaction. In this chapter, we present an overview of the combined stresses with emphasis on drought and bacterial stressors and discuss the stress interaction effect and net effect.

Keywords

Concurrent stress • Stress interaction • New stress • Unique responses

A. Gupta • M. Senthil-Kumar (✉)
National Institute of Plant Genome Research, 10531, JNU Campus, Aruna Asaf Ali Marg, New Delhi 110 067, India
e-mail: skmuthappa@nipgr.ac.in

M. Senthil-Kumar (eds.), *Plant Tolerance to Individual and Concurrent Stresses*, DOI 10.1007/978-81-322-3706-8_1

1.1 Introduction

Under field conditions, the environmental constraints do not always occur independently but most often occur in conjunction with pathogens, and this is detrimental to survival of crop plants. Extreme weather patterns have led to the periodic incidences of drought and pathogen infections (Desprez-Loustau et al. 2007; Yáñez-López et al. 2012; Elad and Pertot 2014). Conventionally, disease triangle represents drought altered plant-pathogen interaction by influencing either the host defence or the pathogen virulence (Achuo et al. 2006; Amtmann et al. 2008; Goel et al. 2008; Hanso and Drenkhan 2009; Atkinson and Urwin 2012). Severe drought in 2003 stimulated *Diplodia pinea* (causal agent of Sphaeropsis blight) epidemic on conifers in Central Europe, and *Diplodia pinea* emerged as a new pathogen infecting *Pinus nigra* in Estonia (Hanso and Drenkhan 2009). The concurrence of drought and pathogen stress and their interaction over plant interface leads to altered plant physiology and resistance responses (Choi et al. 2013; Dossa et al. 2016; Gupta et al. 2016b; Sinha et al. 2016). Plant responses to concurrent drought stress and pathogen infection vary depending on the severity and duration of each stress, nature of infecting pathogens and plant genotype (Achuo et al. 2006; Xu et al. 2008; Ramegowda et al. 2013; Dossa et al. 2016; Gupta et al. 2016a; Sinha et al. 2016). Previously, few studies involving concurrent stresses on plants showed that stress interactions provoke a set of unique plant responses wherein some of the acclimation strategies are attuned to the constraints involved and are not seen under either of the individual stress (Xu et al. 2008; Atkinson et al. 2013; Prasch and Sonnewald 2013a; Gupta et al. 2016b). Moreover, combined stress also evokes responses which are ‘common’ to each of the individual stresses (Prasch and Sonnewald 2013a; Gupta et al. 2016b). Some of these responses are ‘similar’ between combined and individual stress, whereas certain common responses are evoked to a different level under combined stress when compared to individual stress and are termed as ‘tailored responses’. Considering the existence of unique, tailored and similar responses under concurrent drought and pathogen stress compared to individual stresses (Suzuki et al. 2014; Gupta et al. 2016b), it is perceivable that combined stressed plants experience a new state of stress. This underlines the fact that the net impact of a concurrent stress and cognate plant response cannot be studied exclusively from single-stress experiments (Atkinson and Urwin 2012; Suzuki et al. 2014). In this chapter, we attempt to describe the delineation between stress interactions and net impact on plants.

1.2 Stress Interaction

1.2.1 Direct Impact of Drought on Pathogen

Drought stress can influence pathogen survival and spread in environment and thus impacts the disease incidence (Hanso and Drenkhan 2009). In case of rhizosphere-dwelling pathogens, the outcome of interaction between drought stress and bacterial pathogen varies depending upon the nature of the pathogen and whether the pathogen thrives in wet or dry soils. For example, drought favours *Streptomyces scabies* (causal agent of common scab in potato) multiplication in the rhizosphere and thereby increases the opportunity for subsequent infection in plants (Goto 1985). Most often, foliar pathogens experience low water availability in phyllosphere as the most important deterrent. Several studies showed that epiphytic microbial populations increase in wet months (the presence of water) but decline during dry periods (Hirano and Upper 1983, 1990). The leaf surface water does not only support pathogen multiplication but can also be conducive for sustaining plethora of microbial pathogens on phyllosphere and thereby increasing the subsequent chances of plant infection. Ercolani (1991) reported increased diversity of microbial pathogens on olive leaf surface during cooler wet months which however declined during the warmest and driest months of the season. Furthermore, Beattie (2011) reported that abundance of surface water favours bacterial invasion into the leaf tissue. Under water stress, many genes involved in pathogenicity and virulence (of bacteria), including genes in the hypersensitive response and pathogenicity alternative sigma factor (HrpL) regulon, were suppressed in *Pseudomonas syringae* pv. *tomato* DC3000 (causal agent of bacterial speck) (Freeman 2009). Thus, by modulating the pathogen multiplication and survival in the environment, drought (outside the plant) has been shown to regulate incidence of plant infection by these pathogens.

1.2.2 Stress Interactions at the Plant Interface

Drought and pathogen stressors can interact with each other at plant interface wherein drought directly impacts the *in planta* pathogen multiplication and spread. Bacterial movement inside the host is regulated by its flagella, which in turn is favoured by water availability in the leaf apoplast. In an instance, spread of *P. syringae* pv. *syringae* (causes brown spot of beans) in bean seedlings has been shown to be promoted by water (Leben et al. 1970). Instances for drought-induced *in planta* multiplication and spread for other pathogens are also available. Lowered water potential in pea leaves leads to reduced sporulation of *Erysiphe pisi* (causal agent of powdery mildew) (Ayres 1977). Drought stress also restricted *in planta* movement of *Tomato spotted wilt virus* (causal agent of tomato spotted wilt) and attenuated disease symptoms in tomato plants (Córdoba et al. 1991). These examples present a scenario of drought-induced tolerance towards pathogen in plants under combined stress by modulating systemic spread of pathogen.

Drought stress can interfere with plant immunity making the plant susceptible or resistant towards pathogen attack (Mohr and Cahill 2003; Koga et al. 2004; Hatmi et al. 2014). Drought stress imparted susceptibility to *Arabidopsis thaliana* cv. Ler against an avirulent bacterial pathogen *P. syringae* pv. *tomato 1065* (Pst1065) (Mohr and Cahill 2003) and to grapevines against *Xylella fastidiosa* (causal agent of bacterial leaf scorch and Pierce's disease) (Choi et al. 2013). Drought stress increased the severity and progression of leaf scorch disease caused by *X. fastidiosa* in *Parthenocissus quinquefolia* vine (McElrone et al. 2001). The acclimation of *Nicotiana benthamiana* to moderate drought stress reduced the growth of *P. syringae* pv. tabaci (causes wildfire disease in tobacco) (Ramegowda et al. 2013). However, in the same study, severe drought stress had been shown to increase the susceptibility of the plants to *P. syringae*. Here, drought stress increased the ABA accumulation and hence interfered with plant defence responses (Ramegowda et al. 2013).

Water availability facilitated bacterial pathogenesis by suppressing the plant vasculature defences during effector-triggered immunity (ETI) in *A. thaliana* (Cook and Stall 1977; Freeman and Beattie 2009) and PAMP-triggered immunity in *N. benthamiana* (Oh and Collmer 2005). Freeman and Beattie (2009) showed that plants promote ETI and cause localized desiccation at the site of pathogen infection consequently restricting pathogen multiplication. Drought stress tolerance in grapevine involved activation of polyamine oxidation contributing to improved immune response and low susceptibility to *Botrytis cinerea* (causes grey mould disease) (Hatmi et al. 2014).

Drought stress also instigates physiological changes in plants which may be favourable to the pathogen. Drought-stressed sorghum plants were more susceptible to *Macrophomina phaseolina* (causal agent of charcoal rot) infection (Edmunds 1964). Reportedly, the sorghum root volatiles diffuse more rapidly through dry soil and favour *M. phaseolina* infection under drought conditions (Kerr 1964). In another instance, increased *M. phaseolina* infection in drought-stressed common bean has been reported (Mayek-Perez et al. 2002). Ijaz et al. (2013) suggested that drought stress led to accumulation of carbohydrates and amino acids (viz. asparagine and proline) which served as nutrient for the *M. phaseolina* instigating *in planta* pathogen growth and multiplication. Similarly, the drought-induced proline accumulation and ROS metabolism invoked susceptibility towards *Diplodia pinea* in Austrian pine (Sherwood et al. 2015).

In spite of the drought-imposed obstacles for *in planta* pathogen multiplication and survival, pathogen interacts with plant and tends to establish itself *in planta* during combined stress. In an attempt to overcome the obstacle posed by low water availability, bacteria actively modify the leaf surface habitat during drought stress. For example, bacteria can increase the wettability of leaves by secreting surfactants (Bunster et al. 1989; Hutchison and Johnstone 1993). The water films created by these biosurfactants hydrate epiphytic bacterial cells and facilitate movement of bacteria to more favourable sites (Lindow and Brandl 2003). Bacteria also modify their local environment by producing extracellular polymeric substances (EPS) which helps them hold on to the leaf surface and prevent desiccation by encapsulating cells in a hygroscopic matrix (Wilson et al. 1965; Takahashi and Doke 1984). Synthesis of alginate, a component of EPS, is stimulated by desiccation stress in *P. syringae* (Singh et al. 1992; Keith and Bender 1999) and contributes to epiphytic fitness of this organism during drought stress (Yu et al. 1999). High cell densities induce the expression of particular genes (Pierson et al. 1998; Bassler 1999) and contribute to epiphytic fitness (Monier and Lindow 2003) via quorum sensing cell to cell signals. *Xanthomonas campestris* (causal agent of wilt) was able to reverse stomatal closure induced by ABA via secretion of virulence factors (Gudesblat et al. 2009). Taken together, all these evidences suggest that drought influence pathogen multiplication and survival both outside and inside its host. Although at the same time, pathogen has also adopted combat mechanisms and establishes itself in the plant under combined stress conditions.

Further, studies also show that pathogen influences host plant physiology and water relations to predispose it to drought stress. The vascular wilt pathogens cause desiccation state in host plant which leads to reduced photosynthesis and reduced flow of photo assimilates to the roots and eventually causes reduced root growth. As a result, the host plant is more susceptible to the drought stress. *X. fastidiosa*, a xylem-limited bacterial pathogen, induces drought stress in alfalfa (Daugherty et al. 2010). These pathogens colonize and block xylem vessels and reduce their hydraulic conductivity, thereby aggravating the drought stress conditions in plants (Yadeta and Thomma 2013). Tomato plants infected with *Verticillium dahliae* (causal agent of Verticillium wilt) showed decreased leaf water potential (Ayres 1978).

A. thaliana plants infected with *V. longisporum* were tolerant to drought stress. *V. longisporum* induces the expression of vascular-related NAC domain (*VND7*) gene in these plants and triggered de novo xylem formation which leads to enhanced water storage capacity under drought stress conditions (Reusche et al. 2012). *P. syringae* infection in host plant could interfere with plant-water relation by causing water-soaking and the resultant desiccation of the infection site (Beattie 2011), and such case leads to more drought stress experienced by plants.

Both drought stress and foliar bacterial pathogen infection influence ABA levels and stomatal closure in plants. ABA treatment leads to susceptibility of *A. thaliana* towards avirulent bacterium *P. syringae* pv. *tomato* 1065 (Pst1065) infection where the susceptibility increased in a concentration-dependent manner (Mohr and Cahill 2003). Similarly, application of HopAM1 a type III effector of *P. syringae* increases the multiplication and virulence of *P. syringae* under drought stress (Goel et al. 2008). HopAM1 also enhanced ABA-mediated stomatal closure under drought stress (Goel et al. 2008). Pathogen effectors released inside the plant cell cause increased ABA accumulation and stomatal closure and decreased leaf transpiration rate, which altogether improved drought tolerance in combined stressed plants. For instance, application of purified HrpN-a protein produced by *Erwinia amylovora* (causal agent of fire blight) alleviated drought symptoms in *A. thaliana* (Dong et al. 2005). The increased ABA levels in response to the HrpN treatment enhanced the expression of several ABA-signalling regulatory genes as well as the drought-inducible gene *rd29B* (response to dehydration B), the gene product of which mediates ABA-induced responses (Dong et al. 2005). Root colonization with rhizobacteria, *P. chlororaphis* O6 also induced stomatal closure, reduced water loss by transpiration and increased drought tolerance in *A. thaliana* plants (Cho et al. 2008).

In conclusion, the two stressors when co-occurring influence the plant resistance as a result of stress interaction. Thus, in order to study plant-pathogen interaction, the actual scenarios must be accounted, and such understanding cannot be extrapolated from single-stress studies.

1.3 Net Effect of Combined Stresses

From the earlier discussions, it appears that combined stress is perceived as two interacting stressors by the plants where one could see the reminiscence of two individual stresses. However, looking at the existence of a set of unique responses and net impact, it can be settled that combined stress in plants is perceived as a new state of stress. The simultaneous occurrence of more than one stress influences plants as result of stress interaction and direct net impact of occurring stresses together (Daugherty et al. 2010; Atkinson et al. 2013; Ramegowda et al. 2013; Rasmussen et al. 2013; Bostock et al. 2014; Kissoudis et al. 2014; Prasch and Sonnewald 2013a, 2015; Gupta et al. 2016b). The net impact depends on the specific combination of stresses where the concurrence of two stressors can guard or further disrupt plant processes, and both the stresses, when occurring concurrently, most often act in unison to hamper plant growth and development (positive drought-pathogen interaction) (Fig. 1.1). As a result, the combined stresses can cause severe reduction in crop yield when compared with the losses incurred by individual stresses (Siddiqui 1980; Bhatti and Kraft 1992; McElrone et al. 2001; Janda et al. 2008; Prasch and Sonnewald 2013a; Fig. 1.1). Edmunds (1964) observed that concurrent drought stress and *Macrophomina phaseoli* infection caused more damage compared to individual stressed sorghum plants.

The set of net impact resulting from stress interactions in turn depends on common physiological effect or common traits influenced by the two constituent stressors (of concurrent stress) impacting on plant, and the outcome is more devastating than either of the individual stress. Individual drought stress and *X. fastidiosa* infection both lead to low water potential in leaf and influence reduction in stomatal conductance and xylem dysfunction. As a result of such synergism, *X. fastidiosa*, in combination with drought stress, increases the severity and progression of leaf scorch in *Parthenocissus quinquefolia* causing severe reduction in total biomass as compared to individual stresses (Fig. 1.1; McElrone et al. 2001, 2003). Drought stress invokes stomata closure in the plants (Wilkinson and Davies 2002), while on the other hand, *P. syringae* infection signals stomata opening (Melotto et al. 2008). When *Vicia faba* and *A. thaliana* were subjected to a combination *P. syringae* and water deficit, stomatal closure was more pronounced (Ou et al. 2014). In such case of antagonistic stress interaction, responses to abiotic stresses were found to override the responses to biotic stresses (Ou et al. 2014). Recent studies also suggest that the net impact could be the reminiscent of the stress interaction or due to direct impact of combined stress. In the following section, we attempt to delineate and assess the net impact of combined stresses.

1.4 Assessment of Net Impact of Combined Stress

As stated earlier, combined stressed plants experience net impact as a results of one of the following.

(a) Interaction of each stressor with plant
(b) Interaction between two stressor inside the plant
(c) Interaction of one stressor with plant influencing other stressor

In order to tag a combined stressed plant under natural field conditions, dissection of the component contributing towards net impact is important which so far is not understood. Further, for crop protection and improvement, systematic identification of contributory factors (through interaction) to combined drought and pathogen stress is needed. Foremost prerequisite for such studies is the identification of a common agronomically important parameter targeted by both the stressors. For example, yield reduction is seen in case of individual drought and charcoal rot infection in sorghum (Edmunds 1964). The alteration in the identified parameter can be used to answer how the two stressors are interacting with each other and with plant. During such studies, net impact of combined stress can be instanced in one of the following equations:

(i) $CS=D>P$
(ii) $CS=P>D$
(iii) $CS=D+P$ (additive/positive interaction)
(iv) $CS=D\text{-}P$
(v) $CS=P\text{-}D$
(vi) $CS\neq D$ or P or $D+P$ or $D\text{-}p$ or $P\text{-}D$

Here, D, P and CS denote net impact imparted by individual drought and pathogen stresses and their combination, respectively.

In case (i) drought could be said as the 'dominant' stressor, influencing the net impact of the combined stressed plants. The dominant stressor, here, refers to the stress which can modulate the plant processes and decides the plant interaction with subsequent stressor, and also the net impact of the combined stress plants is largely similar to the net impact of dominant stressed plants. In this case, drought can reduce pathogen growth, or it can interfere with plant resistance and impact yield loss (parameter considered here as net impact). For example, drought stress instigated activation of polyamine oxidation and improved immune response which lead to subsequent resistance in grapevine to *Botrytis cinerea* (Hatmi et al. 2014). Likewise, in case (ii) pathogen can be considered as dominant stressor where it can reduce drought effect while interfering with plant-water relations and curb yield loss, e.g. *Erwinia amylovora* alleviated drought symptoms in *A. thaliana* (Dong et al. 2005). In case (iii) the net impact (total loss in yield) is equivalent to the additive losses incurred by individual drought and pathogen stresses and results

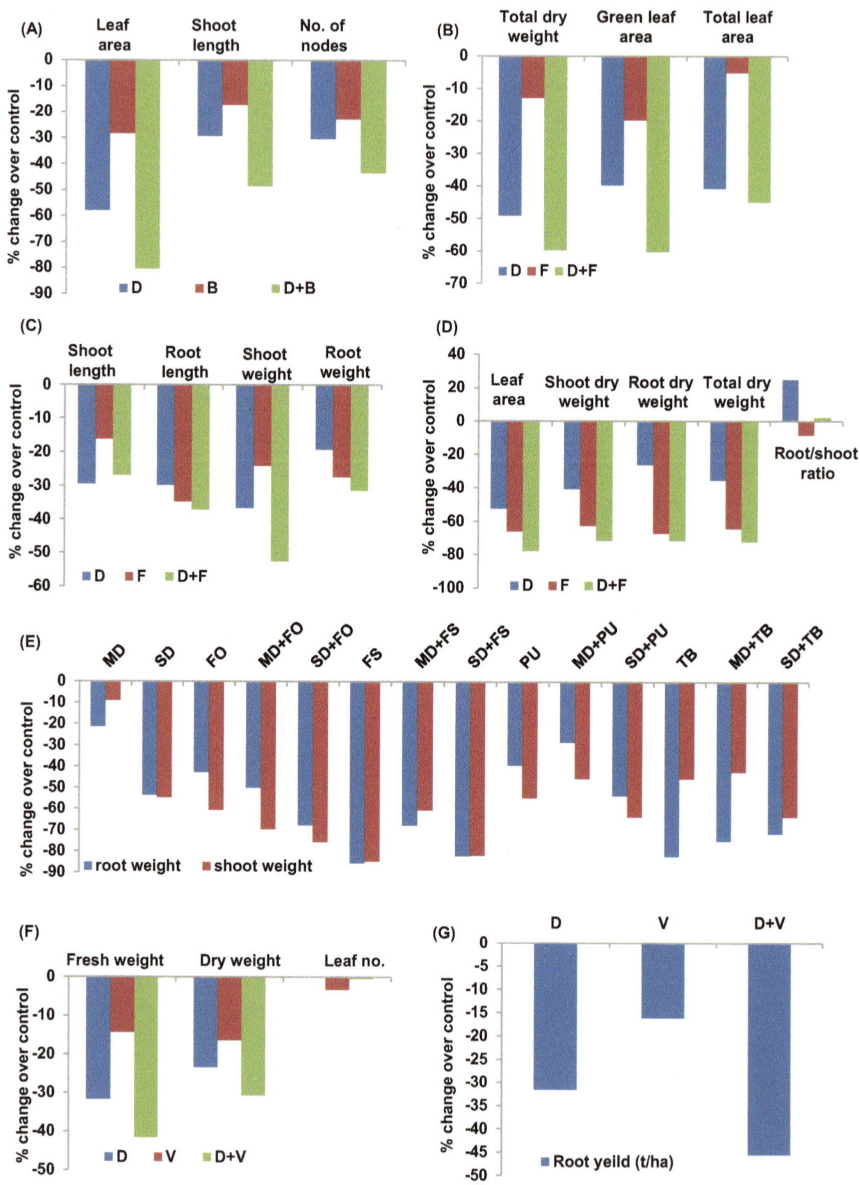

Fig. 1.1 Effects of concurrent drought and pathogen stress on plants. Graphs showing effect of concurrent stresses on yield contributing traits. Drought and bacterial stress (*Xylella fastidiosa*) effect on *Parthenocissus quinquefolia* (McElrone et al. 2001) (**a**), drought and fungal stressor (*Puccinia helianthi*, causes rust) effect on *Helianthus annuus* (Siddiqui 1980) (**b**), drought and fungal (*Drechslera tritici-repentis*) stressor effect on *Triticum aestivum* (Janda et al. 2008) (**c**), drought and fungal (*Macrophomina phaseolina*) stressor effect on *Phaseolus vulgaris* (Mayek-Perez et al. 2002) (**d**), drought and fungal (*Fusarium oxysporum*, FO; *Fusarium solani*,

from positive interaction between drought and pathogen. The example of drought and wilt pathogen interaction can be cited where both these stressors enhance each other's effect in combined stressed plants. In cases (iv) and (v), the net impact, e.g. on yield loss, is equivalent to the subtractive losses (to those incurred by individual stresses) and results mainly because of the antagonistic interaction between two stressors. This instance for this case is presented in Fig. 1.1, where the total biomass under combined stress was the difference of total biomass seen under individual *Pythium ultimum* infection and mild drought stress in *Cicer arietinum* (Bhatti and Kraft 1992). In all these cases (i–v), mainly the stress interaction culminates in net impact which is reminiscent of the individual stresses. In such case, combined stress effect or net impact can be deduced by studying individual stress effect, and shared responses can be exploited for studying stress interaction. However, still another scenario appears where combined stressed plants exhibit a stress effect without stress interaction, e.g. (vi) in this case, the net impact is not equivalent to either of the individual stresses or their positive or negative interaction but is different.

All these situations can be delineated by making use of different stress levels, different pathogen (virulent and avirulent) and different plant genotypes (differing in their resistance responses). By screening different genotypes (exploiting common parameter for stress interaction), one can dissect components of combined stress impact. Recently, a study by Dossa et al. (2016) analysed ten rice genotypes which differed in bacterial blight (BB) resistance (having *R* genes) or drought tolerance (comprising drought quantitative trait loci) or a cross of both BB resistance and susceptible genotype. They imposed different drought stress levels (mild and moderate) and different *Xanthomonas oryzae* pv. oryzae strains (Xoo) (causal agent of rice blight) (virulent PXO99 and avirulent PXO145) under simultaneous stress. Rice genotype IRBB7 (*R* gene, Xa7) showed less Xoo spread and reduced Xoo multiplication under drought stress compared to the well-watered control with PXO145. In contrast, in genotypes with a different BB R gene and/or drought QTLs [IRBB4 (Xa4), IR87705:6.9.B (Xa4 + qDYT2.2), IR87707:445.B.B.B (Xa4 +qDYT2.2 + qDYT4.1) and IR87707:446.B.B.B (Xa4 +qDYT2.2 + qDYT4.1)], Xoo multiplication and spread *in planta* were higher with drought stress. Janda et al. (2008) studied the interaction between drought and fungal stress on wheat. They included three different levels of drought (0, 5, 15 and 20% of PEG) followed by inoculation with *Drechslera tritici*-repentis (DTR, causal agent of tan spot disease

Fig. 1.1 (continued) FS; *Pythium ultimum*, PU; *Thielaviopsis basicola*, TB) stressors effect on *Cicer arietinum* (Bhatti and Kraft 1992) (**e**), drought and viral (*Turnip mosaic virus*) stressor effect on *Arabidopsis thaliana* (Prasch and Somnewald 2013) (**f**) and drought and viral (*beet yellows virus*) stressor on *Beta vulgaris* biomass (Clover et al. 1999) (**g**). The values were extracted from research papers, and % change in yield parameter was calculated over control samples. Negative values in graph denoted reduction in biomass over control treatments. *D* drought, *B* bacteria, *D+B* combined drought and bacterial stress, *F* fungus, *D+F* combined drought and fungal stress, *V* virus, *D+V* combined drought and viral stress, *MD* mild drought stress, *SD* severe drought stress, *FO Fusarium oxysporum, FS Fusarium solani, PU Pythium ultimum, TB Thielaviopsis basicola*

in wheat) at two different time (6 and 72 h after the PEG treatment) in different wheat genotypes with two DTR resistant (M-3 and Mv Magvas) and two sensitive (Bezostaya 1 and Glenlea) varieties. While 15% PEG reduced the level of infection in sensitive Bezostaya variety, 20% PEG treatment lowered the tolerance level of M-3.

Both these studies indicated drought as the dominant stressor, where it might have affected the plant resistance in influencing the *in planta* pathogen multiplication. On similar lines, Prasch and Somnewald attempted to study the natural genetic variation of combined biotic and drought stress response, by studying the expression profile of common genes (between individual and combined stress) in natural accessions of Arabidopsis (Prasch and Sonnewald 2013b).

1.5 Combined Stress as a New State of Stress: Reminiscent and Different from Either of the Individual Stresses

Certain physiological responses are modulated in a plant under combined stress which is either unique or common (tailored or similar with individual stressed plants). For instance, concurrent viral and drought-stressed plants accumulated proline at a level different from individual drought stress or viral infection (Xu et al. 2008). Moreover, these combined stressed plants did not accumulate sucrose which was induced upon individual virus infection. Ascorbic acid content in drought-stressed plants declined by 37.5% and was undetectable in virus-infected plants. However, concurrent stressed plants did not show any change in ascorbic acid levels as compared to control plants. Likewise, more anthocyanins were accumulated in concurrent stressed plants over individual stressed plants. The above-mentioned instance explains existence of common responses between individual and combined stresses which are tailored to suit the plant defences under combined stress.

The extent of common (tailored or similar) and unique responses between combined and individual stresses depends on the nature of pathogens that infect drought-stressed plants. Transcriptome studies in *A. thaliana* exposed to concurrent drought stress and *P. syringae* infection revealed that 31% of the differentially expressed transcripts were unique to concurrent dual stresses and were lacking under individual stress treatments and 22% were common with either of the individual stress (Gupta et al. 2016b). However, under *X. fastidiosa* and drought stress interaction, 56% of differentially regulated genes were shared with either of the individual stresses (Choi et al. 2013). Gupta et al. (2016b) studied transcriptome profile of *A. thaliana* and compared time of occurrence of pathogen during concurrent drought and *P. syringae* infection. Comparison of differentially expressed genes across individual and combined stress drought and *P. syringae* infection revealed 505 genes unique to drought followed by pathogen stress and 885 unique genes under pathogen followed by drought combined stress (Gupta et al. 2016b). The existence of common genes between individual and combined stresses indicates that plants economize their defence resources while using existing stress-responsive molecular machinery for upcoming new or additional stresses.

Thus, although the biotic and abiotic stress response pathways share common responses, the net effect of concurrent abiotic and biotic stress interaction on plants cannot be predicted from the individual stressed plants (Suzuki et al. 2014). Reports indicates that these common responses can also be tailored in terms of the magnitude or fold change which cannot be extrapolated from individual stress response (Atkinson and Urwin 2012; Prasch and Sonnewald 2015; Rasmussen et al. 2013) (Fig. 1.1). Based on these evidences, we propose that the concurrent stress combinations are perceived by plants as a 'new stress' leading to a reprogramming of the defence responses while compared to plants under individual stress. In addition to these different responses, combined stressed plants also maintain a state reminiscent of individual stresses.

1.6 Conclusions and Future Perspective

Frequent incidences of combined drought and pathogen stress result in inevitable losses in crop yields. The limited understanding on the plant responses towards combined stress highlights the importance of nature of infecting pathogen, time of occurrence of each stress, intensity of the stress, plant age and genotype of the plant. The two constituent stressors of combined stress can interact with each other outside or inside the plant and influence plant resistance and physiology. Each stressor can interact with plant genotype and modulate physiology and resistance response towards subsequent stressor. In these cases, the net impact on plant can be predicted from independent stress studies. Alternatively, the combined stress can directly be perceived as a different stress, and the resultant net impact is not reminiscent of either of the individual stresses. In such case, the net impact on the plant is difficult to comprehend from individual stress studies and warrant an explicit study to dissect the combined stress responses. In this purview, studying the plant genetic architecture with reference to combined stress is a viable approach. Incorporating different levels of stress, time and plant genotype in future studies helps to dissect the constituents of combined stress responses, while at the same time, the need for identification of a parameter to screen combined stressed plants is of utmost importance. Altogether, the gained knowledge can be better translated to assess the utilization and environmental risks of different genotypes under combined stress. The increased understanding of plant responses and genetic architecture under combined stress further opens up avenues for breeding programmes for improvement of cultivars.

Acknowledgement Combined stress tolerance-related projects at MS-K Lab are supported by the National Institute of Plant Genome Research core funding, DBT-Ramalingaswami re-entry fellowship grant (BT/RLF/re- entry/23/2012) and DBT-innovative young biotechnologist award. AG acknowledges the SERB-National Post-Doctoral Fellowship (N-PDF/2015/000116).

References

Achuo EA, Prinsen E, Höfte M (2006) Influence of drought, salt stress and abscisic acid on the resistance of tomato to *Botrytis cinerea* and *Oidium neolycopersici*. Plant Pathol 55:178–186

Amtmann A, Troufflard S, Armengaud P (2008) The effect of potassium nutrition on pest and disease resistance in plants. Physiol Plant 133:682–691

Atkinson NJ, Urwin PE (2012) The interaction of plant biotic and abiotic stresses: from genes to the field. J Exp Bot 63:3523–3543

Atkinson NJ, Lilley CJ, Urwin PE (2013) Identification of genes involved in the response of Arabidopsis to simultaneous biotic and abiotic stresses. Plant Physiol 162:2028–2041

Ayres PG (1977) Effects of leaf water potential on sporulation of *Erysiphe pisi* (pea mildew). Trans Br Mycol Soc 68:97–100

Ayres PG (1978) Water relations of diseased plants. In: Kozlowski TT (eds) Water deficits and plant growth, vol V. Academic Press, London, pp 1–60

Bassler BL (1999) How bacteria talk to each other: regulation of gene expression by quorum sensing. Curr Opin Microbiol 2:582–587

Beattie GA (2011) Water relations in the interaction of foliar bacterial pathogens with plants. Annu Rev Phytopathol 49:533–555

Bhatti MA, Kraft JM (1992) Influence of soil moisture on root rot and wilt of chickpea. Plant Dis 76:1259–1262

Bostock RM, Pye MF, Roubtsova TV (2014) Predisposition in plant disease: exploiting the nexus in abiotic and biotic stress perception and response. Annu Rev Phytopathol 52:17–49

Bunster L, Fokkema NJ, Schippers B (1989) Effect of surface-active *Pseudomonas* spp. on leaf wettability. Appl Environ Microbiol 55:1340–1345

Cho SM, Kang BR, Han SH, Anderson AJ, Park JY, Lee YH, Cho BH, Yang KY, Ryu CM, Kim YC (2008) 2R,3R-butanediol, a bacterial volatile produced by *Pseudomonas chlororaphis* O6, is involved in induction of systemic tolerance to drought in *Arabidopsis thaliana*. Mol Plant-Microbe Interact 21:1067–1075

Choi H-K, Iandolino A, Goes da Silva F, Cook D (2013). Water deficit modulates the response of *Vitis vinifera* to the Pierce's disease pathogen *Xylella fastidiosa*. Mol Plant-Microbe Interact 26:643–657

Clover GRG, Smith HG, Azam-Ali SN, Jaggard KW (1999) The effects of drought on sugar beet growth in isolation and in combination with beet yellows virus infection. J Agric Sci (Camb) 133:251–261

Cook AA, Stall RE (1977) Effects of watersoaking on response to *Xanthomonas vesicatoria* in pepper leaves. Phytopathol. 67:1101–1103

Córdoba AR, Taleisnik E, Brunotto M, Racca R (1991) Mitigation of tomato spotted wilt virus infection and symptom expression by water stress. J Phytopathol 133:255–263

Daugherty M, Lopes JS, Almeida RP (2010) Strain-specific alfalfa water stress induced by *Xylella fastidiosa*. Eur J Plant Pathol 127:333–340

Desprez-Loustau ML, Robin C, Reynaud G, Deque M, Badeau V, Piou D, Hussone C, Marçaise B (2007) Simulating the effects of a climate-change scenario on the geographical range and activity of forest-pathogenic fungi. Can J Plant Pathol 29:101–120

Dong HP, Yu H, Bao Z, Dong H (2005) The ABI2-dependent abscisic acid signalling controls HrpN-induced drought tolerance in Arabidopsis. Planta 221:313–327

Dossa GS, Torres R, Henry A, Oliva R, Maiss E, Cruz CV, Wydra K (2016) Rice response to simultaneous bacterial blight and drought stress during compatible and incompatible interactions. Eur J Plant Pathol. doi:10.1007/s10658-016-0985-8

Edmunds LK (1964) Combined relation of plant maturity, temperature and soil moisture to charcoal stalk rot development in grain sorghum. Phytopathology 54:514–517

Elad Y, Pertot I (2014) Climate change impacts on plant pathogens and plant diseases. J Crop Improv 28:99–139

Ercolani GL (1991) Distribution of epiphytic bacteria on olive leaves and the influence of leaf age and sampling time. Microb Ecol 21:35–48

Freeman BC (2009) The role of water stress in plant disease resistance and the impact of water stress on the global transcriptome and survival mechanisms of the phytopathogen *Pseudomonas syringae*. PhD thesis, Iowa State University, 114 pp; 3355217

Freeman BC, Beattie GA (2009) Bacterial growth restriction during host resistance to *Pseudomonas syringae* is associated with leaf water loss and localized cessation of vascular activity in *Arabidopsis thaliana*. Mol Plant-Microbe Interact 22:857–867

Goel AK, Lundberg D, Torres MA, Matthews R, Akimoto-Tomiyama C, Farmer L, Dangl JL, Grant SR (2008) The *Pseudomonas syringae* type III effector HopAM1 enhances virulence on water stressed plants. Mol Plant-Microbe Interact 21:361–370

Goto K (1985) The relative importance of precipitation and sugar content in potato peel for the detection of the incidence of common scab (*Streptomyces scabies*). Soil Sci Plant Nutr 31:419–425

Gudesblat GE, Torres PS, Vojnov AA (2009) Xanthomonas campestris overcomes Arabidopsis stomatal innate immunity through a DSF cell-to-cell signal-regulated virulence factor. Plant Physiol 149:1017–1027

Gupta A, Dixit SK, Senthil-Kumar M (2016a) Drought stress predominantly endures *Arabidopsis thaliana* to *Pseudomonas syringae* infection. Front Plant Sci 7:808. doi:10.3389/fpls.2016.00808

Gupta A, Sarkar AK, Senthil-Kumar M (2016b) Global transcriptional analysis reveals unique and shared responses in *Arabidopsis thaliana* exposed to combined drought and pathogen stress. Front Plant Sci 7:686

Hanso M, Drenkhan R (2009) *Diplodia pinea* is a new pathogen on Austrian pine (*Pinus nigra*) in Estonia. New Dis Rep 19:14

Hatmi S, Gruau C, Trotel-Aziz P, Villaume S, Rabenoelina F, Baillieul F, Eullaffroy P, Clément C, Ferchichi A, Aziz A (2014) Drought stress tolerance in grapevine involves activation of polyamine oxidation contributing to improved immune response and low susceptibility to *Botrytis cinerea*. J Exp Bot 66:775–787

Hirano SS, Upper CD (1983) Ecology and epidemiology of foliar bacterial plant pathogens. Annu Rev Phytopathol 21:243–269

Hirano SS, Upper CD (1990) Population biology and epidemiology of *Pseudomonas syringae*. Annu Rev Phytopathol 28:155–177

Hutchison ML, Johnstone K (1993) Evidence for the involvement of the surface active properties of the extracellular toxin tolaasin in the manifestation of brown blotch disease symptoms by *Pseudomonas tolaasii* on *Agaricus bisporus*. Physiol Mol Plant Pathol 42:373–384

Ijaz S, Sadaqat HA, Khan AN (2013) A review of the impact of charcoal rot (*Macrophomina phaseolina*) on sunflower. J Agric Sci 151:222–227

Janda T, Cséplõ M, Németh CS, Vida GY, Pogány M, Szalai G, Veisz O (2008) Combined effect of water stress and infection with the necrotrophic fungal pathogen *Drechslera tritici-repentis* on growth and antioxidant activity in wheat. Cereal Res Commun 36:53–64

Keith LM, Bender CL (1999) AlgT (sigma22) controls alginate production and tolerance to environmental stress in *Pseudomonas syringae*. J Bacteriol 181:7176–7184

Kerr A (1964) The influence of soil moisture on infection of peas by *Pythium ultimum*. Aust J Biol Sci 17:676–685

Kissoudis C, van de Wiel C, Visser RGF, van der Linden G (2014) Enhancing crop resilience to combined abiotic and biotic stress through the dissection of physiological and molecular cross talk. Front Plant Sci 5:207

Koga H, Dohi K, Mori M (2004) Abscisic acid and low temperatures suppress the whole plant-specific resistance reaction of rice plants to the infection of *Magnaporthe grisea*. Physiol Mol Plant Pathol 65:3–9

Leben C, Schroth MN, Hildebrand DC (1970) Colonization and movement of *Pseudomonas syringae* on healthy bean seedlings. Phytopathology 60:677–680

Lindow SE, Brandl MT (2003) Microbiology of the phyllosphere. Appl Environ Microbiol 69:1875–1883

Mayek-Perez N, Garcia-Espinosa R, Lopez-Castaneda C, Acosta-Gallegos JA, Simpson J (2002) Water relations, histopathology and growth of common bean (*Phaseolus vulgaris* L.) during pathogenesis of *Macrophomina phaseolina* under drought stress. Physiol Mol Plant Pathol 60:185–195

McElrone AJ, Sherald JL, Forseth IN (2001) Effects of water stress on symptomatology and growth of *Parthenocissus quinquefolia* infected by *Xylella fastidiosa*. Plant Dis 85:1160–1164

McElrone AJ, Sherald JL, Forseth IN (2003) Interactive effects of water stress and xylem-limited bacterial infection on the water relations of a host vine. J Exp Bot 54:419–430

Melotto M, Underwood W, He SY (2008) Role of stomata in plant innate immunity and foliar bacterial diseases. Annu Rev Phytopathol 46:101–122

Mohr PG, Cahill DM (2003) Abscisic acid influences the susceptibility of *Arabidopsis thaliana* to *Pseudomonas syringae* pv. tomato and *Peronospora parasitica*. Funct Plant Biol 30:461–469

Monier JM, Lindow SE (2003) Differential survival of solitary and aggregated bacterial cells promotes aggregate formation on leaf surfaces. Proc Natl Acad Sci USA 100:15977–15982

Oh H-S, Collmer A (2005) Basal resistance against bacteria in *Nicotiana benthamiana* leaves is accompanied by reduced vascular staining and suppressed by multiple *Pseudomonas syringae* type III secretion system effector proteins. Plant J 44:348–359

Ou X, Gan Y, Chen P, Qiu M, Jiang K, Wang G (2014) Stomata prioritize their responses to multiple biotic and abiotic signal inputs. PLoS One 9(7):e101587

Pierson LS 3rd, Wood DW, Pierson EA (1998) Homoserine lactone-mediated gene regulation in plant-associated bacteria. Annu Rev Phytopathol 36:207–225

Prasch CM, Sonnewald U (2013a) Simultaneous application of heat, drought, and virus to Arabidopsis plants reveals significant shifts in signaling networks. Plant Physiol 162:1849–1866

Prasch CM, Sonnewald U (2013b) In silico selection of *Arabidopsis thaliana* ecotypes with enhanced stress tolerance. Plant Signal Behav 8(11):e26364

Prasch CM, Sonnewald U (2015) Signaling events in plants: stress factors in combination change the picture. Environ Exp Bot 114:4–14

Ramegowda V, Senthil-Kumar M, Ishiga Y, Kaundal A, Udayakumar M, Mysore KS (2013) Drought stress acclimation imparts tolerance to *Sclerotinia sclerotiorum* and *Pseudomonas syringae* in *Nicotiana benthamiana*. Int J Mol Sci 14:9497–9513

Rasmussen S, Barah P, Suarez-Rodriguez MC, Bressendorff S, Friis P, Costantino P, Bones AM, Nielsen HB, Mundy J (2013) Transcriptome responses to combinations of stresses in Arabidopsis. Plant Physiol 161:1783–1794

Reusche M, Thole K, Janz D, Truskina J, Rindfleisch S, Drübert C, Polle A, Lipka V, Teichmann T (2012) Verticillium infection triggers VASCULAR-RELATED NAC DOMAIN7-dependent de novo xylem formation and enhances drought tolerance in Arabidopsis. Plant Cell 24:3823–3837

Sherwood P, Villari C, Capretti P, Bonello P (2015) Mechanisms of induced susceptibility to Diplodia tip blight in drought-stressed Austrian pine. Tree Physiol 35:549–562

Siddiqui MQ (1980) Some effects of rust infection and moisture stress on growth, diffusive resistance and distribution pattern of labelled assimilates in sunflower. Aust J Agric Res 31:719–726

Singh S, Koehler B, Fett WF (1992) Effect of osmolarity and dehydration on alginate production by fluorescent pseudomonads. Curr Microbiol 25:335–339

Sinha R, Gupta A, Senthil-Kumar M (2016) Understanding the impact of drought on foliar and xylem invading bacterial pathogen stress in chickpea. Front Plant Sci 7:902. doi:10.3389/fpls.2016.00902

Suzuki N, Rivero RM, Shulaev V, Blumwald E (2014) Mittler R Tansley review Abiotic and biotic stress combinations. New Phytol 203:32–43

Takahashi T, Doke N (1984) A role of extracellular polysaccharides of *Xanthomonas campestris* pv. cirri in bacterial adhesion to citrus leaf tissues in preinfectious stage. Jpn J Phytopathol 50:565–573

Wilkinson S, Davies WJ (2002) ABA-based chemical signalling: the co-ordination of responses to stress in plants. Plant Cell Environ 25:195–210

Wilson HA, Lilly VG, Leach JG (1965) Bacterial polysaccharides IV. Longevity of *Xanthomonas phaseoli* and *Serratia marcescens* in bacterial exudates. Phytopathology 55:1135–1138

Yadeta KAJ, Thomma BP (2013) The xylem as battleground for plant hosts and vascular wilt pathogens. Front Plant Sci 4:97. doi:10.3389/fpls.2013.00097

Yáñez-López R, Torres-Pacheco I, Guevara-González RG, Hernández-Zul MI, Quijano-Carranza JA, Rico-García E (2012) The effect of climate change on plant diseases. Afr J Biotechnol 11:2417–2428

Yu J, Penaloza-Vazquez A, Chakrabarty AM, Bender CL (1999) Involvement of the exopolysaccharide alginate in the virulence and epiphytic fitness of *Pseudomonas syringae pv. syringae*. Mol Microbiol 33:712–720

Xu P, Chen F, Mannas JP, Feldman T, Sumner LW, Roossinck MJ (2008) Virus infection improves drought tolerance. New Phytol 180:911–921

Closing the Biotic and Abiotic Stress-Mediated Yield Gap in Cotton by Improving Soil Management and Agronomic Practices

2

Gunasekhar Nachimuthu and Ashley A. Webb

Abstract

Intensive agricultural practices in conjunction with climate change in the recent past have resulted in outbreaks of abiotic and biotic stresses that pose challenges to modern cotton farming systems around the world. Even with improved transgenic cotton varieties, the average lint yield realised in developing (India) and developed countries (Australia) is about 500 and 2500 kg/ha, respectively, compared with theoretical potential yield of 5000 kg/ha. The yield gap is largely associated with many factors being out of balance in the soil and crop management and climate that induce these biotic and abiotic stresses which impacts on the yield. Filling this yield gap requires a joint venture among various agricultural disciplines that include agronomy, soil science, physiology and molecular biology. Several major research projects have aimed to increase yield, and they are related to management of stress and development of stress-tolerant cotton varieties. Bt cotton and herbicide-tolerant cotton are example outcomes from research conducted to alleviate biotic stress. This review briefly describes the major abiotic and biotic stresses in cotton production. Thereafter, the role of soil and agronomic practices in stress management is outlined. This chapter covers drought stress, temperature stress and the major pathogenic stresses and provides appropriate management strategies. This review will be useful broadly

G. Nachimuthu (✉)
New South Wales Department of Primary Industries, Australian Cotton Research Institute, 21888 Kamilaroi Highway, Narrabri, NSW 2390, Australia
e-mail: guna.nachimuthu@dpi.nsw.gov.au

A.A. Webb
New South Wales Department of Primary Industries, Tamworth Agricultural Institute, 4 Marsden Park Rd, Calala, NSW 2340, Australia

M. Senthil-Kumar (eds.), *Plant Tolerance to Individual and Concurrent Stresses*,
DOI 10.1007/978-81-322-3706-8_2

to the plant science community, especially physiologists and molecular biologists who will be encouraged to design their research projects based on field realities, considering soil characteristics and agronomic practices.

Keywords
Combined stress • Agricultural production • Yield gap • Cotton agronomy • Drought • Pathogen

2.1 Introduction

Intensive agricultural practices to meet the food and fibre demands of a rapidly rising global population (FAO 2002) have led to large-scale environmental problems, especially the degradation of soil and water resources (Bunzel et al. 2015; Drinkwater and Snapp 2007; Hart et al. 2004; Nachimuthu et al. 2016). This, coupled with extreme weather events induced by climate change, is directly impacting crop growth and productivity due to the occurrence of multiple stresses in field crops. The major causal factors of these stresses include temperature (heat and cold), water (drought and water logging), soil biology (pathogens), soil chemical constraints (salinity, nutrient limitation) and soil physical constraints (soil compaction due to heavy machinery or soil dispersion induced by sodicity). These causal factors can be classified as biotic (pathogens, weeds, pest) or abiotic (temperature, drought etc.) stresses. Abiotic stresses could cause up to a 50% reduction in yield (Boyer 1982). Several major research projects have aimed to manage these stresses and thereby improve yields by developing stress-tolerant cotton varieties, while the plants themselves evolve various molecular and physiological adaptations to protect themselves (Pandey et al. 2015). Development of genetically modified (GM) cotton and its high rate of adoption are evidence of its success. However, soil and agronomic management of crops is equally important in alleviating those stresses. The synergy between genetic and agronomic management is core to productivity improvement and sustainability. It is hard to be predictive about breeding by agronomy synergies, as they could result from unanticipated advances in one field or the other; so it would be unwise to assume they have been exhausted (Fischer 2009). In this chapter, though we have selected cotton as a model crop to enumerate the biotic and abiotic stresses and the role of soil and crop management in their mitigation, we also briefly cover other crops in general and identify the major reviews published on biotic and abiotic stresses, their occurrence, management and plant responses.

2.2 Cotton Growth

A successful crop relies on optimal climatic conditions and effective management practices. Cotton is an indeterminate crop, and in its native habitat as a perennial shrub, it can survive for several years. Under conditions of stress, such as low soil moisture, excessive heat or nutrient deficiency, cotton plants often adapt by shedding some of their squares, flowers or bolls to guarantee survival with constrained resources (Williams and Bange 2015). Cotton yields are often reported to decrease under stressful conditions induced by limitations of root growth (Karamanos et al. 2004). However, being an indeterminate crop, cotton can often compensate for poor growing conditions by having an extended period of growth, although this might impact fibre quality where a mechanised single harvest system is being practiced in developed countries, such as Australia. Some regions in Australia, such as Southern New South Wales where recent cotton production area is increasing (NSW DPI 2013), often have a narrow window for cotton planting and harvest. This is due to seasonal temperatures and climate in that region (Bange 2004) and means that any stress during the crop often impacts the maturity time and subsequent yield.

2.3 Closing the Stress-Mediated Yield Gap

Agricultural intensification to boost higher yields and achieve frequent harvests not only demands very high inputs but can also result in off-farm environmental impacts across the world (Nachimuthu et al. 2016). Two options available to produce enough food and fibre for the growing population include intensification of existing land use or land clearing and expansion of agricultural land. Land clearing and farmland expansion are not feasible in several countries due to a lack of suitable land for crop production as well as government policies and regulations aimed at conserving remnant native vegetation to provide a range of ecosystem services. Further, with the current climate change debate, land clearing could form a major source of greenhouse gas emissions and could have negative consequences for global communities. Therefore, the most feasible option available is the sustainable intensification of existing cropping lands together with the reduction of environmental consequences. This strategy is successfully proven in the Australian cotton industry, where the cotton yield is >2.5 times the world average and where the industry is well recognised for its social and environmental licence to operate (Nachimuthu and Webb 2016). There has been a 40% improvement in water-use productivity (lint yield/ML of water) of Australian cotton achieved by the improvement of cultivars, genetic modification to manage biotic stress (weeds and pest) and implementation of best management practices (Nachimuthu and Webb 2016; Roth et al. 2013). The number of sprays dropped significantly by 89% after the introduction of GM cotton, thereby improving farming system sustainability (Cotton Australia 2016). In addition, in-season cultivation to eradicate weeds has largely ceased, reducing machine traffic and soil compaction in the field. Therefore, genetic improvements have played a major role in alleviating biotic (cotton pest and weeds) and abiotic stresses (water use productivity improvement). In spite of this improvement, there is still a large yield gap in Australian cotton. The average cotton yield under irrigated systems in Australia is around 2500 kg lint/ha compared to the potential yield of 5000 kg/ha (Constable and Bange 2015). The yield of cotton in India, even after the introduction of GM cotton, is only around 500 kg lint/ha

(Blaise et al. 2014), and a large yield gap still exists. The yield gap varies across countries, and the yield potential is not a fixed factor as it could be improved by a combination of genetics and management (Constable and Bange 2015). The major yield gap factors include biotic (pests, diseases and weeds) and abiotic stresses induced by water limitation, soil constraints such as compaction, sodicity, nutrient deficiency and climate (extreme temperatures). Filling this yield gap requires a joint venture among various agricultural disciplines that include agronomy, soil science, physiology and molecular biology.

In the following sections, we briefly describe the biotic stresses (Sect. 2.4) and abiotic stresses (Sects. 2.5 and 2.6) in cotton and the role of soil and crop management to evade or alleviate the intensity of those stresses. Some of the other published reviews have covered the occurrence, management and physiological response of plants to various stresses. They include, but are not limited to, abiotic and biotic stress combinations described in detail by Suzuki et al. (2014) and plant response to cross-tolerance between abiotic and biotic stress combinations described in detail by Rejeb et al. (2014).

2.4 Biotic Stress in Cotton and Agronomic Management Responses

Biotic stresses in cotton farming systems include pests, fungi, virus, bacteria and weeds. Temperature, humidity and rainfall are the key factors that control the incidence of pests and diseases. At present, bollworm (*Helicoverpa* sp.) and weeds are not a major issue in those cotton farming systems where Bollgard® and Roundup Ready Flex® varieties are used (Mahmood ur et al. 2014). However, a new challenge is emerging in the form of Roundup Ready volunteer cotton, and an herbicide-resistant weed could pose a potential problem in the future.

Apart from bollworm, some important bacterial and fungal diseases of cotton include root rot (caused by *Rhizoctonia bataticola* and *R. solani* and *Pythium* spp.), black root rot (caused by *Thielaviopsis* spp.), blight (caused by *Phoma exigua*) and wilt (caused by *Fusarium* spp.), leaf spot (caused by *Alternaria macrospora*) and bacterial blight (caused by *Xanthomonas* sp.). Different weather conditions like temperature, rainfall, and soil types and plant physiological status influence the severity of the diseases. For example, wilt diseases in cotton are reported to occur more severely on sandy soils and under wet seasons. Low temperatures and high humidity favour the development of blight caused by *P. exigua* (Koenning 2016). On the other hand, chilling stress enhances the susceptibility of cotton plants to *A. macrospora* (Zhao et al. 2012). Similarly, it has been found that high moisture and low temperature conditions increase *R. solani* infections, while warmer temperatures and low soil moisture favour the infection of *R. bataticola* in cotton plants (Monga and Raj 2016). The disease control measures for the seedling diseases like root rot and stem canker, in general, involve the use of fungicides like mefenoxam and iprodione along with cultural agronomic practices that make conditions less conducive for the pathogens (Koenning 2016). Crop rotations in cotton farming systems can have both positive and negative effects on biotic stress depending on the rotation crop chosen (Kirby and Smith 2016; Nachimuthu 2016). Selection of suitable rotation crops to break the pest and disease cycle is a vital strategy in integrated disease management. Research on disease suppressive microbes is ongoing although the success of inoculum establishment in soil is still a challenge (Pereg and McMillan 2015).

2.5 Drought Stress in Cotton and Agronomic Management

Drought is one of the most important abiotic stress factors limiting cotton growth and production around the world. The occurrence of drought-impacted cropping fields is expected to double in this century (Deeba et al. 2012), and the competition for water between urban, industrial and agricultural use will compound the drought effects. Plants acclimatise to drought or water deficit using a range of measures. The first response in several crops is stomatal closure (Tombesi et al. 2015). The detailed impact of drought on plants and their physiological response is covered in detail by Farooq et al. (2009). Cotton plants possess a range of morphological, physiological, biochemical and anatomical traits that are tolerant of, and adaptive to, drought. These include, but are not limited to, a deep root system, high root-shoot ratio, a decrease in leaf area expansion, epicuticular waxiness, leaf rolling, maintenance of leaf turgidity at lower water potential, osmoregulation, an increase in stomata per unit leaf area, stomatal regulation-closure of stomata, a decrease in transpiration rate in response to stomatal closure, early senescence and cutout, increased abscission of fruiting parts, an increase in proline content, increase in nonprotein amino acids, cellular adaptations such as increase in cell wall thickness, development of mechanical tissue and a decrease in cell size and intercellular space (CICR 2016). In general, the above drought tolerance or resistance mechanisms (Farooq et al. 2009) can be grouped into:

1. Morphological – which includes drought escape-reduced life cycle, drought avoidance-stomatal control and altered root-shoot ratio
2. Physiological and molecular mechanisms

A significant untapped opportunity to overcome drought stress exists by exploring, understanding and exploiting the soil conditions which enhance resource use by plant roots (Whitmore and Whalley 2009). Commercially grown cotton species (*Gossypium hirsutum*) always respond positively to water supply when irrigated compared to growth patterns under dryland conditions. The difference in yield between rain grown and irrigated cotton in Australia in the past 5 years is a typical example of the impact of drought stress on cotton (Table 2.1). Yield losses of 16–43% due to water stress have been reported in cotton (Constable and Hearn 1981). According to the Australian Cotton Production Manual 2015 (Williams and Bange 2015), the preferred agronomic measure practiced by cotton growers to alleviate drought stress is to alter the row configuration (Payero et al. 2012). Adopting single-

Table 2.1 Australian cotton yield estimates

Year	Irrigated Yield (bales/ha)	Dryland Yield (bales/ha)
2009/10	9.4	3.6
2010/11	9.2	2.8
2011/12	10.0	3.3
2012/13	10.8	3.6
2013/14	10.2	1.6
2014/15	12.1	2.7

Source: Cotton Year Book (2015)
1 bale = 227 kg lint

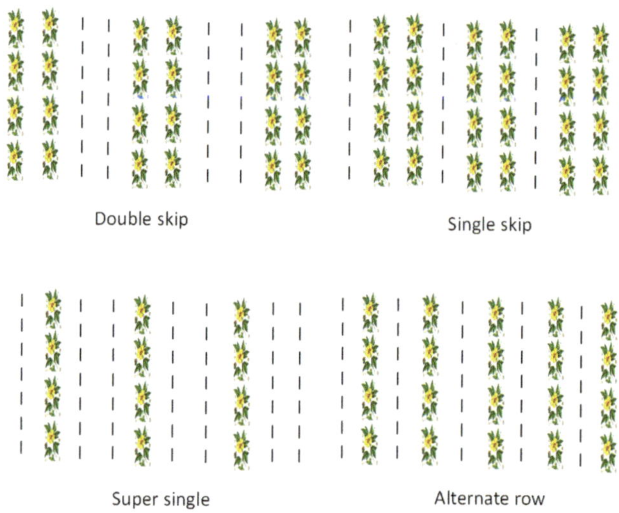

Fig. 2.1 Diagrammatic representation of model row configurations to overcome water-limited situations. *Dashed line* represents rows without cotton plants (Pendergast et al. 2015)

skip and double-skip row configurations will reduce the plant population and yield in years with good rainfall and soil moisture compared with solid row planting (Fig. 2.1), yet it is a risk management strategy to manage the effects of drought. Single-skip row configuration is recommended for soils with high water holding capacity, and double-skip configuration is recommended for soils with low water holding capacity (Bange 2015). Other agronomic management practices such as crop rotation with cereal crops could enhance soil health and subsequently improve infiltration, thereby enhancing soil water storage and availability for the subsequent crop (Hulugalle and Scott 2008). Surface stubble retention of rotation cereal crops is often recommended to improve soil structure and the establishment of healthy seedlings in the early stage (Soil Quality 2016). Plant physiologists and molecular scientists working under controlled environments need to account for these factors in the field environment when developing new tools to manage drought stress.

The scientific evidence suggests the best way to manage drought and enhance productivity is through judicious use of two basic natural resources: soil and water (Gardner and Gardner 1983; Ostle et al. 2009). Drought management in cotton needs special care, although there is a lack of solutions available for rain-fed cotton showing wilting symptoms when no rain is forecast in the short term. Some agronomic practices such as cultivation could be avoided to prevent aggravating the situation. Even a minimum amount of tillage is capable of causing root damage, and any cultivation for weed control needs to be avoided or delayed, where possible. In addition, surface fertiliser dressing could cause leaf-burnt symptoms leading to additional stress (Rochester 2001b). Application of plant growth regulators such as Pix® needs to be carefully managed under situations of drought stress (Albers and Schnakenberg 1994).

2.6 Temperature Stress in Cotton

Climate change is predicted to cause higher than the average global temperatures (Luo et al. 2016). Temperature plays a key role in the growth and development of all crops. The rate of cotton plant growth is largely determined by temperature (Ullah et al. 2016). Though cotton plants are suited to hot climatic conditions, extreme temperatures can lead to lower yields (Ullah et al. 2016). Both cold (<15 °C) and hot (>36 °C) temperatures can delay the development of cotton crops. In the United States (USA) and Australia, the accumulation of heat units termed as day degrees (DD) is used to measure the development of cotton where:

$$DD = (\text{maximum temperature} - 12 + \text{minimum temperature} - 12)/2 \text{ (Kirkham 2014).}$$

If the minimum temperature is less than 12, then DD= (maximum temperature – 12)/2 (Williams and Bange 2015).

Cotton plants often receive cold shock when the temperature drops below 11 °C, and this can delay the crop development (Bange and Milroy 2006). The DD is critical for several cotton developmental stages. Examples of target DD for various cotton growth stages are presented in the Australian cotton production manual (Williams and Bange 2015). Plant responses to cold stress and physiological responses to cold tolerance mechanisms are covered in Yadav (2010), whereas for cotton production under a changing climate, we have focused mainly on heat or high temperature stress.

A negative relationship has been observed between high temperatures and cotton lint yield in Arkansas in the USA (Oosterhuis 1999). Temperature stress in cotton directly affects its vegetative and reproductive biology by influencing the number of vegetative and reproductive branches of the plant. Cotton plants are inclined to produce a higher proportion of vegetative branches and a lower proportion of reproductive or fruiting branches under high temperatures (Reddy et al. 1991). In particular, there are additional heat stress-induced effects on cotton reproductive development. The time between each phase of reproductive maturation can be affected which, in turn, impacts the yield. For example, first-square initiation, first flowering and first boll formation and opening are decreased with higher than normal temperatures. The boll and square retention stages are largely decreased under high temperatures (Reddy et al. 1999).

Cotton is a strong absorber of solar radiation and the temperature of cotton plant tissue will be higher on clear sunny days. Some plants adopt a strategy of having waxy surfaces to reflect the radiation; however, cotton plants tend to absorb radiation and this aggravates the drought stress. Thick cuticles and hairiness are desirable characteristics to minimise the heat stress effects, and varieties with those characteristics could reduce the impact of heat stress (CICR 2016). There is a trend of higher night temperatures recorded in recent years (BOM 2016), and this could also adversely affect the cotton plants' response to heat stress. In the night time, cotton plants close their stomata, and evaporative cooling through leaves does not occur.

High temperatures could directly affect the growth and yield of cotton and indirectly induce drought stress by high evaporation demand leading to concurrent stress. While irrigation is effective in enabling crops to overcome water stress, dryland cotton will not only experience water stress but also the concurrent stress resulting from higher temperatures. The ability of the cotton plant to overcome both temperature and concurrent drought stress is largely dependent on the ability of its roots to adapt to those concurrent stresses. Plants under temperature stress on a dry soil could be limited by the flow of water into the plant and out the leaves. The soil water content in those situations is less important than the ability of the soil and/or plant roots to translocate the water into the plant. There are new plant-based irrigation scheduling methods being developed using canopy temperature sensors (White and Raine 2008), and the technological change will play a crucial role in alleviating crop stress. Canopy sensor-based irrigation scheduling could address the limitations of accessibility to water by plant roots as a result of other soil physical constraints that may not be possible using capacitance probes or neutron probes which sense water content in soil. Cotton growing in sandy soils often wilts during the afternoon, despite good soil moisture, due to the poor movement of water in sandy soils after free drainage. Research conducted under controlled environments such as small pots often ignores such factors. In view of this, we discuss plant roots and their impact on abiotic stress in the next section.

2.7 Effect of Root Morphology and Response on Abiotic Stresses

Drought management strategies could be either plant based or soil management based. Understanding and exploiting the soil conditions in which the plant roots are able to maximise their resource use efficiency could have significant potential in overcoming drought and temperature stresses. A detailed review of the physical effects of soil drying on roots and crop growth was presented by Whitmore and

Whalley (2009). The root morphology along with its density in topsoil and subsoil tends to influence the way the plants could manage these stresses. Plant roots which have the potential to proliferate deeper into the soil could assist plants in overcoming drought stress (Ho et al. 2005). Recent research in Australia on root morphology highlights the fine root production of Bollgard Roundup Ready ® (GM) cotton is much lower than conventional cotton (Hulugalle et al. 2015). Another research project focusing on phosphorus acquisition suggested that cotton plants might be sourcing phosphorus from deep soil, and there is a lack of response in the topsoil (Bell 2015). This suggests either the roots are inactive in the topsoil or the plant roots are evolved or adopted to source the nutrients from deep soil under water-limited conditions. This water-limited drought situation has led to another stress-inducing subsoil P depletion cum nutrient stratification, and this was clearly documented in grain cropping soils of Australia (Bell et al. 2012), a similar soil type to where cotton is grown (vertisols).

Detailed reviews have been undertaken on root-shoot signaling that controls the shoot elongation and stomatal function; however, it is yet to be resolved how this signal operates under concurrent stress situations of combined biotic and abiotic factors and their interactions (Whitmore and Whalley 2009).

Some researchers assume that if a soil is used in controlled experiments in a glasshouse or growth cabinets, the results of the experiments can be extrapolated to field conditions. However, such results should be applied with caution as there could be a multitude of errors associated with this logic if the interpretation is inappropriate. Some of the simple reasons could include that the plant responses are typical of a particular soil type and most of the physical properties of soil (such as bulk density, root exploration volume, lack of drainage etc.) (Håkansson and Lipiec 2000) in the controlled environment are likely to be different from the field conditions. The main factor to consider is that the root density in the pots will be higher than under field conditions, depending on pot size, and this will have a direct effect on the results (Poorter et al. 2012).

2.8 Effect of Soil and Crop Management on Negating the Impact of Biotic and Abiotic Stresses

Both biotic and abiotic stress factors could lead to several crop management challenges associated with pests, diseases and poor nutrient supply and uptake by plants. Drought-induced soil changes could increase cracking, surface crusting and soil structure degradation and aggregation. Depending on soil clay content and the type of clay minerals present, the depth of the cracking could vary more than a metre deep. Soil crusting could become an issue influencing soil hydrology if there is a lack of organic matter in the soil (Overstreet and DeJong-Huges 2016). These changes in soil structure can negatively affect the soil-plant-water relationships and reduce water use efficiency (Loch et al. 2005), in particular leading to poor root system development which in turn reduces the above-ground biomass (Wrona et al. 1999). Studies on root elongation of cotton as a function of soil strength and soil water content have shown that root elongation is more affected by soil strength than by soil moisture (Taylor and Ratliff 1969).

There are several management practices available to alleviate the negative effects of drought and heat stresses under field conditions. They include, but are not limited to, soil management practices, irrigation, crop residue management and the selection of crop varieties. Soil tillage could affect surface soil moisture and temperature or heat balance. In particular, evaporation and infiltration are directly impacted by tillage as the heat exchange between the atmosphere and the soil is altered by tillage. Previous research suggests that compared to conventional tilled fields, soils with a history of no or minimum tillage are characterised by a greater number of longitudinally continuous bio-pores (as a result of enhanced soil biology and better plant root development) (Dighton et al. 1997). Minimum tillage or zero tillage often provides better soil conditions for undisturbed root growth as there will be less mechanical disturbance of soil structure (Fageria and Moreira 2011). This will have a direct influence on soil moisture conservation and infiltration and subsequently a negating effect on drought stress. In situations where a subsoil hardpan or compaction has occurred as a result of the use of heavy machinery, the rooting depth in those soils could be enhanced by deep ripping or deep tillage. This operation could consume a high amount of energy but is essential to alleviate the soil condition, so it is able to overcome the concurrent stresses.

Crop residue cover plays a significant role in soil sustainability (Farooq et al. 2011). Surface mulching of crop residues reduces the soil temperature as a result of low heat conductivity of mulch compared to soil. Surface mulch also prevents soil evaporation and conserves soil water and enhances the formation of soil aggregates (Hulugalle et al. 2014). An additional benefit of surface residues includes physical protection of the soil from potential erosion during heavy storm events. Therefore, stubble retention practices could alter the soil water balance and play a significant role in alleviating drought stress. Cotton uses very limited water from sowing to first flowering (Bange et al. 2005), and planning a sowing window to capture the seasonal rainfall is a key factor to avoid stress during critical crop growth stages.

Soil health can impact the intensity of drought and temperature stress effects on the plants. Transpiration rate as a result of plant stomatal conductance was reported to be lower under high fertile soil compared to low fertile soil (Irmak 2016). Applications of plant nutrients such as potassium could influence the effects of temperature stress by modifying the stomatal function and improving temperature stress tolerance (Oosterhuis et al. 2014). The role of potassium in influencing the physiological response of cotton and other crops to various stresses (drought stress, cold stress, salt stress and biotic stress) was described in detail by Oosterhuis et al. (2014). Micronutrients or trace elements could influence the growth of plants by modifying biochemical attributes, and antioxidant defence mechanisms and their management are suggested as important factors in stress tolerance (Habibi 2014).

The Australian cotton industry has now adopted (>95%) stubble mulching practices against the old practice of raking and burning cotton stubbles. This practice has been proven to enhance the cycling of nutrients and improve soil health (Rochester 2011) and could mitigate drought stress to some extent. However, under some abiotic stress situations such as pathogens, burning of stubble is practiced to overcome the disease, although the beneficial effect of burning on disease control in cotton is disputed. The reason being, the senescence of cotton leaves occurs during the boll opening stage and the disease inoculum is already spread across the field. Raking and burning might be ineffective in those situations (Rochester 2001a). The best management practice to reduce the inoculum causing seedling diseases (e.g. *Pythium*) in those situations could be to reduce the amount of stubble from cotton or rotation crops on the soil surface. Fusarium inoculum could be better controlled by retaining crop residues on the soil surface for the longest duration possible before incorporation. The management of biotic and abiotic stresses requires the development of stress-tolerant cultivars in conjunction with the soil and crop management practices described above.

2.9 Conclusion and Research Gaps

Crop production under field conditions can be decreased by several abiotic and biotic stresses. Under field situations, crops are exposed to multiple stresses. Studies covering multiple stresses and causal factors, rather than a single stress, could provide tangible solutions. Plant reactions to a combination of drought and heat stress cannot be directly extrapolated from the response of plants to each of these different stresses when applied individually (Mittler 2006). Co-occurrence of heat and drought stress affects plants to a larger degree than the cumulative effects of those individual stresses occurring at different points in time. Soil and crop management play a vital role in negating the effect of a single stress or concurrent multiple stresses. Plant physiologists and molecular biologists need to consider these effects while planning their research. There is relatively little information available about the effects of drought and heat stresses along with other soil constraints such as soil compaction, erosion, salinity and acidification. The main aspect to consider while investigating multiple stress combinations is to address them as a new state of stress in plants and not simply the sum of two different stresses (Mittler 2006). We propose that future research requires a joint venture among various agricultural disciplines that include agronomy, soil science, physiology and molecular biology. This will bring together complementary skills and expertise to solve a common problem by utilising a constructive and collaborative approach. In other words, to overcome concurrent biotic and abiotic stresses, interdisciplinary teams of researchers working at the field scale, cellular and plant level must collaborate to deliver effective solutions. It is also important to study the effect of the concurrent abiotic stresses on plant pathogen interactions at the different levels starting from inoculum propagation, pathogen infection and disease establishment in order to be able to devise strategies to enhance the tolerance of plants to the combined biotic and abiotic stresses.

Acknowledgements We acknowledge Senthil-Kumar Muthappa and book chapter reviewers for their valuable feedback and suggestions.

References

Albers DW, Schnakenberg CT (1994) Plant growth regulators for cotton. Web source: http://extension.missouri.edu/p/G4258; Downloaded 15 July 2016

Bange MP (2004) The impact of temperature extremes on cotton performance. Final report. Cotton Research and Development Corporation, Australia, 27

Bange MP (2015) Raingrown (dryland) cotton. Australian cotton production manual 2015. Chapter 4, pp 20–24

Bange M, Milroy SP (2006) Impact of cold shock on early cotton plant development. The Australian Cotton Grower, June–July 2006:33–35

Bange MP, Carberry PS, Marshall J, Milroy SP (2005) Row configuration as a tool for managing rain-fed cotton systems: review and simulation analysis. Aust J Exp Agric 45:65–77

Bell M (2015) Developing soil testing and fertiliser response guidelines to manage P, K and S fertility for irrigated and dryland cotton cropping systems, Final report to Cotton Research and Development Corporation. Queensland Alliance for Agriculture and Food Innovation, The University of Queensland, Gatton

Bell M, Lester D, Smith L, Want P (2012) Increasing complexity in nutrient management on clay soils in the northern grain belt – nutrient stratification and multiple nutrient limitations. In: 16th Australian agronomy conference, 14–18th October 2012, University of New England in Armidale, NSW

Blaise D, Venugopalan MV, Raju AR (2014) Introduction of Bt cotton hybrids in India: did it change the agronomy? Indian J Agron 59:1–20

BOM. Bureau of Meteorology. Commonwealth of Australia (2016) http://www.bom.gov.au/

Boyer JS (1982) Plant productivity and environment. Science 218:443–448

Bunzel K, Schäfer RB, Thrän D, Kattwinkel M (2015) Pesticide runoff from energy crops: A threat to aquatic invertebrates? Sci Total Environ 537:187–196

CICR (2016) Abiotic stresses in cotton – a physiological approach. Central Institute for Cotton Research

Constable GA, Bange MP (2015) The yield potential of cotton (*Gossypium hirsutum* L.). Field Crop Res 182:98–106

Constable GA, Hearn AB (1981) Irrigation for crops in a sub-humid environment. Irrig Sci 3:17–28

Cotton Australia (2016) Biotechnology and cotton. Cotton Australia- Cotton library fact sheets.

Cotton Year Book (2015) The Australian cottongrower- Annual cotton year book 2015, vol 36. PO box 766, Toowoomba 4350.

Deeba F, Pandey AK, Ranjan S, Mishra A, Singh R, Sharma YK et al (2012) Physiological and proteomic responses of cotton (*Gossypium herbaceum* L.) to drought stress. Plant Physiol Biochem 53:6–18

Dighton J, Jones HE, Robinson CH, Beckett J (1997) The role of abiotic factors, cultivation practices and soil fauna in the dispersal of genetically modified microorganisms in soils. Appl Soil Ecol 5:109–131

Drinkwater LE, Snapp SS (2007) Nutrients in agroecosystems: rethinking the management paradigm. Adv Agron 92:163–196

Fageria NK, Moreira A (2011) Chapter four – the role of mineral nutrition on root growth of crop plants. In: Donald LS (ed) Advances in agronomy, vol 110. Academic Press, London, pp 251–331

FAO (2002) World agriculture: towards 2015/2030. Economic and Social Development Department of FAO, Rome

Farooq M, Wahid A, Kobayashi N (2009) Plant drought stress: effects, mechanisms and management. Agron Sustain Dev 29:185–212

Farooq M, Flower KC, Jabran K, Wahid A, Siddique KHM (2011) Crop yield and weed management in rainfed conservation agriculture. Soil Tillage Res 117:172–183

Fischer RA (2009) Chapter 2 – farming systems of Australia: exploiting the synergy between genetic improvement and agronomy. In: Crop physiology. Academic Press, San Diego, pp 22–54

Gardner WR, Gardner HR (1983) Plant production and management under drought conditions. Principles of water management under drought conditions. Agric Water Manag 7:143–155

Habibi G (2014) Chapter 14 – role of trace elements in alleviating environmental stress. In: Emerging technologies and management of crop stress tolerance. Academic Press, San Diego, pp 313–342

Håkansson I, Lipiec J (2000) A review of the usefulness of relative bulk density values in studies of soil structure and compaction. Soil Tillage Res 53:71–85

Hart MR, Quin BF, Nguyen ML (2004) Phosphorus runoff from agricultural land and direct fertilizer effects. J Environ Qual 33:1954–1972

Ho MD, Rosas JC, Brown KM, Lynch JP (2005) Root architectural tradeoffs for water and phosphorus acquisition. Funct Plant Biol 32:737–748

Hulugalle NR, Scott F (2008) A review of the changes in soil quality and profitability accomplished by sowing rotation crops after cotton in Australian vertosols from 1970 to 2006. Soil Res 46:173–190

Hulugalle NR, Heimoana V, Kimber S, Powell J (2014) Managing carbon in cotton based farming systems- Final report submitted to Cotton Research and Development Corporation. NSW Department of Primary Industries, Narrabri, p. 125

Hulugalle NR, Broughton KJ, Tan DKY (2015) Fine root production and mortality in irrigated cotton, maize and sorghum sown in vertisols of northern New South Wales, Australia. Soil and Tillage Res 146(Part B):313–322

Irmak S (2016) Impacts of extreme heat stress and increased soil temperature on plant growth and development. Web material- University of Nebraska-Lincoln; http://cropwatch.unl.edu/2016/impacts-extreme-heat-stress-and-increased-soil-temperature-plant-growth-and-development

Karamanos AJ, Bilalis D, Sidiras N (2004) Effects of reduced tillage and fertilization practices on soil characteristics, plant water status, growth and yield of upland cotton. J Agron Crop Sci 190:262–276

Kirby K, Smith L (2016) Tackling verticilium. Spotlight Magazine, Cotton Research and Development Corporation. Winter 2016:16–18

Kirkham MB (2014) Chapter 27 – stress-degree-day concept and crop water stress index. In: Principles of soil and plant water relations, 2nd edn. Academic Press, Boston, pp 483–499

Koenning S (2016) Cotton stem canker, wet weather blight, or ascochyta blight. Cotton disease information note No. 2. Accessed on 26 July 2016

Loch RJ, Grant CG, McKenzie DC, Raine SR (2005) Improving plants' water use efficiency and potential impacts from soil structure change – research investment opportunities, Final report to the National Program for Sustainable Irrigation. CRCIF report number 3.14/1. Cooperative Research Centre for Irrigation Futures, Toowoomba

Luo Q, Bange M, Braunack M, Johnston D (2016) Effectiveness of agronomic practices in dealing with climate change impacts in the Australian cotton industry – a simulation study. Agric Syst 147:1–9

Mahmood ur R, Qasim M, Bukhari SA, Shaheen T (2014) Chapter 6 – Bt crops: a sustainable approach towards biotic stress tolerance. In: Emerging technologies and management of crop stress tolerance. Academic Press, San Diego, pp 125–142

Mittler R (2006) Abiotic stress, the field environment and stress combination. Trends Plant Sci 11:15–19

Monga D, Raj S (2016) Root rot disease of cotton and its management. CICR technical bulletin no: 3. Central Institute for Cotton Research, Nagpur. Accessed 26 July 2016

Nachimuthu G (2016) Crop rotation and soil health. Spotlight Magazine. Cotton Research and Development Corporation. Winter 2016:19

Nachimuthu G, Webb AA (2016) On-farm soil conservation measures in cotton farming systems of Australia: a sustainability analysis. J Soil Water Conserv 71:75A–80A
Nachimuthu G, Halpin NV, Bell MJ (2016) Effect of sugarcane cropping systems on herbicide losses in surface runoff. Sci Total Environ 557–558:773–784
NSW DPI (2013) Moving in and out of cotton – identifying farming systems issues in southern NSW irrigation areas (Proposal 1). New South Wales Department of Primary Industries and Cotton Research and Development Corporation. http://www.insidecotton.com/xmlui/bitstream/handle/1/4247/DAN1309%20Final%0Report_%20Southern%20NSW.pdf?sequence=3&is Allowed=y
Oosterhuis DM (1999) Yield response to environmental extremes in cotton. In: Oosterhuis DM (ed) Proceedings of the 1999 cotton research meeting and summaries of cotton research in progress. Arkansas Agricultural Experiment Station special report, 193, pp 30–38
Oosterhuis DM, Loka DA, Kawakami EM, Pettigrew WT (2014) Chapter three – the physiology of potassium in crop production. In: Donald LS (ed) Advances in agronomy, vol 126. Academic Press, London, pp 203–233
Ostle NJ, Levy PE, Evans CD, Smith P (2009) UK land use and soil carbon sequestration. Land Use Policy 26(Suppl 1):S274–S283
Overstreet LF, DeJong-Huges J (2016) The importance of soil organic matter in cropping systems of the Northern Great Plains. University of Minnesota- Extension 2016; Web material- http://www.extension.umn.edu/agriculture/tillage/importance-of-soil-organic-matter/
Pandey P, Ramegowda V, Senthil-Kumar M (2015) Shared and unique responses of plants to multiple individual stresses and stress combinations: physiological and molecular mechanisms. Front Plant Sci 6
Payero J, Robinson G, Harris G, Singh D (2012) Water extraction of solid and skip-row cotton. In: 16th Australian agronomy conference 14–18th October 2012, University of New England in Armidale, NSW
Pendergast L, Harris G, Grabham M, Purcell J, Montgomery J (2015) Irrigated or semi-irrigated cotton. Australian cotton production manual 2015; Chapter 5. pp 24–28
Pereg L, McMillan M (2015) Scoping the potential uses of beneficial microorganisms for increasing productivity in cotton cropping systems. Soil Biol Biochem 80:349–358
Poorter H, Bühler J, van Dusschoten D, Climent J, Postma JA (2012) Pot size matters: a meta-analysis of the effects of rooting volume on plant growth. Funct Plant Biol 39:839–850
Reddy VR, Reddy KR, Baker DN (1991) Temperature effect on growth and development of cotton during the fruiting period. Agron J 83:211–217
Reddy KR, Davidonis GH, Johnson AS, Vinyard BT (1999) Temperature regime and carbon dioxide enrichment alter cotton boll development and fiber properties. Contribution from the Dep. of Plant and Soil Sciences, Mississippi State Univ., and the USDA-ARS Southern Regional Res. Ctr., New Orleans, LA. Mississippi Agric. and Forestry Exp. Stn. Paper no. J9391. Agron J 91:851–858
Rejeb I, Pastor V, Mauch-Mani B (2014) Plant responses to simultaneous biotic and abiotic stress: molecular mechanisms. Plants 3:458
Rochester I (2001a) Cotton stubble management. NutriPak- a practical guide to cotton nutrition; Chapter 13
Rochester I (2001b) NITROGEN. NutriPak – a practical guide to cotton nutrition; Chapter 2
Rochester IJ (2011) Sequestering carbon in minimum-tilled clay soils used for irrigated cotton and grain production. Soil Tillage Res 112:1–7
Roth G, Harris G, Gillies M, Montgomery J, Wigginton D (2013) Water-use efficiency and productivity trends in Australian irrigated cotton: a review. Crop Pasture Sci 64:1033–1048
Soil Quality (2016) Benefits of retaining stubble — New South Wales. Soil Quality Pty Ltd. http://www.soilquality.org.au/factsheets/benefits-of-retaining-stubble-nsw
Suzuki N, Rivero RM, Shulaev V, Blumwald E, Mittler R (2014) Abiotic and biotic stress combinations. New Phytol 203:32–43
Taylor HM, Ratliff LF (1969) Root elongation rates of cotton and peanuts as a function of soil strength and soil water content. Soil Sci 108:113–119

Tombesi S, Nardini A, Frioni T, Soccolini M, Zadra C, Farinelli D et al (2015) Stomatal closure is induced by hydraulic signals and maintained by ABA in drought-stressed grapevine. Sci Report 5:12449

Ullah K, Khan N, Usman Z, Ullah R, Saleem FY, Shah SAI et al (2016) Impact of temperature on yield and related traits in cotton genotypes. J Integr Agric 15:678–683

White S, Raine SR (2008) A grower guide to plant based sensing for irrigation scheduling. National Centre for Engineering in Agriculture Publication, 1001574/6: USQ, Toowoomba

Whitmore AP, Whalley WR (2009) Physical effects of soil drying on roots and crop growth. J Exp Bot 60:2845–2857

Williams S, Bange MP (2015) The cotton plant. Australian cotton production manual 2015; Chapter 1, pp 8–10

Wrona AF, Oosterhuis DM, McMichael B (1999) Getting to the root of your crop's health. Cotton Physiol Today 10:1–8

Yadav SK (2010) Cold stress tolerance mechanisms in plants. A review. Agron Sustain Dev 30:515–527

Zhao J, Li S, Jiang T, Liu Z, Zhang W, Jian G et al (2012) Chilling stress—the key predisposing factor for causing *Alternaria alternata* infection and leading to cotton (*Gossypium hirsutum* L.) leaf senescence. PLoS One 7:e36126

Impact of Concurrent Weed or Herbicide Stress with Other Biotic and Abiotic Stressors on Crop Production

3

Muthukumar Bagavathiannan, Vijay Singh, Prabhu Govindasamy, Seth Bernard Abugho, and Rui Liu

Abstract

Crop plants are exposed to several biotic and abiotic stress factors under natural production conditions. These factors may interact with each other, leading to altered impacts on crop yield. Weeds and herbicides often represent important stress factors on crops. Their interaction with other biotic (insects, diseases, nematodes, etc.) and abiotic (temperature, moisture, etc.) stress factors may influence the nature and dynamics of crop-weed interactions and thereby impact crop yields. In many cases, interaction of weeds/herbicides with other stress factors can lead to accelerated crop yield loss. Yet, such interactions were not well studied and fully understood; existing research knowledge in this area is widely scattered in the literature. This paper establishes, by reviewing existing literature, a foundation of information on the outcomes of interactions between weeds/herbicides and other stressors. This literature analysis also illustrates the need for more targeted research on this topic, which is becoming increasingly important on the face of global climate change.

Keywords

Weed • Herbicide • Combined stress • Disease • Drought • Temperature stress • Crop yield • Agronomic practices • Potential yield

M. Bagavathiannan (✉) • V. Singh • P. Govindasamy • S.B. Abugho • R. Liu
Department of Soil and Crop Sciences, Texas A&M University, 370 Olsen Blvd, 77843-2474, College Station, TX, USA
e-mail: muthu@tamu.edu

M. Senthil-Kumar (eds.), *Plant Tolerance to Individual and Concurrent Stresses*,
DOI 10.1007/978-81-322-3706-8_3

3.1 Introduction

Changes to plant physiological processes due to climatic variabilities may induce stress (Patterson 1995; Teixeira et al. 2013), restricting a plant from achieving its potential growth (Wheeler et al. 2000). Abiotic stress is caused by physical forces (e.g., light, temperature, humidity, greenhouse gases), whereas biotic stress is caused by biological organisms (e.g., weeds, insects, diseases, nematodes) to which a plant is exposed during its growth period. Several studies have highlighted the negative impact on crop growth of extreme temperature conditions (Wheeler et al. 2000; Peng et al. 2004; Prasad et al. 2006a, b), moisture stress (Manikavelu et al. 2006; Steduto et al. 2012), relative humidity (Jia et al. 2015), weeds (Barrentine 1974; Murphy et al. 1996), insects (Oerke 2006; Seiter et al. 2013), and diseases (Mundt et al. 1995; Obilo et al. 2010). However, simultaneous occurrence of two or more of these stresses in a given environment may alter the overall outcome on crop productivity.

Weeds are known to be the most serious pest of crops, which can incur severe yield penalties when left uncontrolled (Bloomberg et al. 1982; Hall et al. 1992). Severe weed infestations can even lead to complete crop failure. The degree of weed stress on crops depends upon the competitive ability of a crop through its various physiological and morphological attributes that allow the crop plant to utilize light, water, space, and nutrients efficiently in the presence of a weed. The competitive interactions of weeds on crops can be influenced tremendously by conditions such as temperature, moisture, CO_2, insects, and diseases (e.g., Patterson 1995; Alberto et al. 1996). As a result, simultaneous occurrence of weeds and other biotic and abiotic stress conditions can have an accelerated impact on crop yield. For example, Flint and Patterson (1983) found that weed infestations combined with high-temperature stress may drastically reduce soybean yield.

In addition to weeds, stress induced by herbicide applications may also interact with biotic and abiotic stressors in impacting crop yields. Herbicides are widely used as an effective management tool to control weeds. Tolerant crop varieties typically have mechanisms to metabolize herbicides (at varying degrees) applied onto them. As a result, herbicides may alter the physiology of crops temporarily, making them susceptible to insect and disease infection (Bradley et al. 2002; Duke et al. 2007). Adequate consideration to the combined effect of herbicide stress and other biotic and abiotic stress factors is vital.

The combined effect of multiple stress factors involving weeds and/or herbicides has not been well studied. The focus of this review is to establish a foundation, based on existing research findings, on the impact of combined stress involving weeds/herbicides on crop productivity.

3.2 Stress Combinations

3.2.1 Impact of Combined Weed/Herbicide and Temperature Stress on Crops

Temperature is an important driving force for plant development (Evans 1993; Roberts et al. 1996; Hatfield and Prueger 2015). Suboptimal temperature conditions (low or high temperatures) can influence several plant physiological processes and affect normal plant growth and development (Wheeler et al. 2000; Teixeira et al. 2013; Barlow et al. 2015). High temperature/heat stress occurring for even few hours can drastically reduce crop yields (Prasad et al. 2000; Porter and Semenov 2005). High-temperature stress is more severe when it occurs at critical crop development stages, especially the reproductive stage (Mesihovic et al. 2016). A long-term yield analysis (1979–2003) has revealed the reduction in rice grain yield by 10% for every 1 °C increase in night temperatures in Manila, Philippines (Peng et al. 2004). Low temperature/cold stress also affects crop production. Cold stress reduces seed germination, delays phenological development, and increases flower sterility, all of which can impact crop yields (Allen and Ort 2001; Rymen et al. 2007). Nie et al. (1992), for example, reported that exposing maize seedlings to 15 °C or below could seriously reduce photosynthetic activity of the leaves. Yield reduction due to temperature stress is crop specific and follows different physiological mechanisms that limit the formation of sinks for photosynthates (Teixeira et al. 2013).

Temperature fluctuations not only affect crops but also influence the phenological development of weeds and thereby alter the nature of crop-weed interactions (Patterson 1995). Flint et al. (1983) have shown in a replacement series experiment that cotton experienced more competition from spurred anoda (*Anoda cristata* (L.) Schlecht.) at day/night temperatures of 26°/17 °C, compared with 32°/23 °C. Potter and Jones (1977) reported that the ratio of relative growth rate of common cocklebur (*Xanthium strumatrium*) to soybean declined with increasing temperatures. High-temperature conditions are favorable for C_4 plants since they require higher optimum temperatures for photosynthesis, compared to that of C_3 plants (Yamori et al. 2014). Further, C_4 plants show improved metabolic activity and growth under heat stress compared to C_3 plants under heat stress (Tiaz and Zeiger 1991). As a result, C_4 weeds can compete well with C_3 crops under high-temperature/heat stress conditions.

Herbicide applications also cause stress on crops. When herbicide applications coincide with suboptimal temperature conditions, the effects on crop growth can be magnified. Natural tolerance of crops to herbicides is typically achieved by metabolic detoxification of the compounds by the crop. Herbicide injury on a tolerant crop is usually less when applied to rapidly growing, actively metabolizing plants that are free from environmental stress. Temperature variations, however, can influence the uptake and metabolism of herbicides (Muzik 1976; Dickson et al. 1990; Godar et al. 2015). Low-temperature stress following herbicide applications may reduce crop growth and increase herbicide injury due to reduced metabolic activity. Thompson et al. (1970) observed herbicide injury on corn at some locations in Illinois due to postemergence applications of atrazine under cold stress. It was found that low-temperature conditions decreased the rate of detoxification of foliar-absorbed atrazine, which caused high crop injury. Likewise, Viger et al. (1991) found that injury to corn from *S*-metolachlor was greater when it was grown under low-temperature conditions. Higher crop injury following herbicide applications under cold conditions may also enhance the competitive interaction of weeds on crops and reduce yields.

High temperature/heat stress, on the other hand, reduces the absorption and translocation of herbicides due to stomatal closure and rapid drying of the herbicide droplet on the leaf surface (Legg 1983). Further, the leaves of plants grown at high temperatures produce more epicuticular wax, which reduces herbicide retention and absorption (Price 1983; Pillmoor 1985). Harrison and Peterson (1999) reported that phytotoxicity of preemergence (PRE) and postemergence (POST) applications of oxyfluorfen on broccoli (*Brassica oleracea* L. var. *italica*) cultivars declined at higher temperatures (25/20 °C), compared to that of lower temperatures (15/10 °C). Although high-temperature conditions may reduce potential herbicide injury to crops, they may also reduce weed control efficacy, leading to more weed escape, interference, and crop yield loss. For instance, fenoxaprop-ethyl activity on wild oat (*Avena fatua* L.) was low under high-temperature conditions (Xie et al. 1996). Palmer amaranth (*Amaranthus palmeri* S. Watson) was less sensitive to mesotrione at high temperatures (40/30 °C), but control greatly increased at low temperatures of 25/15 °C (Godar et al. 2015). Ge et al. (2011) observed that subjecting glyphosate-resistant plants to low temperatures could make them sensitive to glyphosate.

In certain cases, however, herbicide absorption, translocation, and efficacy can be high under high temperatures (Masiunas and Weller 1988; Matzenbacher et al. 2014). The absorption and translocation of glyphosate in johnsongrass (*Sorghum halepense* (L.) Pers.) and Bermuda grass (*Cynodon dactylon* (L.) Pers.) increased as temperature increased after treatment (Jordan 1977). Fausey and Renner (2001) reported that efficacy of flumiclorac on *Chenopodium album* and *Amaranthus retroflexus* increased as temperature increased from 10 °C to 40 °C. Similarly, control of *Ipomoea lacunosa* was greater when acifluorfen was applied at 35/26 °C (day/night), compared to 27/18 °C (Lee and Oliver 1982). The combined effect of weed pressure, herbicide applications, and temperature stress on crop growth and development is complex and not well studied. More research is thus imperative in this area.

3.2.2 Impact of Combined Weed/Herbicide and Moisture Stress on Crops

Soil moisture and crop production are intricately associated. Soil moisture has always been one of the main factors limiting crop production in much of the world where rainfall is erratic and insufficient for crop growth (Kramer 1980; Steduto et al. 2012). Moisture stress occurs when the loss of water through transpiration exceeds water absorption (Jaleel et al. 2009; Blum 2011). Moisture stress lowers stomatal cell turgor, which decreases gas exchange, leaf elongation, and CO_2 uptake (Kirnak et al. 2001). Drought stress during reproductive stage can result in >80% yield reduction in maize (Monneveux et al. 2006; Kamara et al. 2003), 70% in chickpea (Nayyar et al. 2006), 60% in sunflower (Mazahery-Laghab et al. 2003), and 70% in soybean (Samarah et al. 2006). In addition to drought stress, excessive moisture/flooding stress can also have a detrimental impact on crop yield. The degree of crop yield reduction in response to drought/flooding depends on crop genotype and other interacting factors.

Soil moisture stress directly affects weed physiology, including seed germination, growth, and weed-crop interactions. C_4 weed species have higher tolerance for moisture stress compared with C_3 crops (Ozturk et al. 1981). Parminder et al. (2015) found that weeds with deep taproot system [e.g., Palmer amaranth] can exhibit more interference by extracting moisture from much deeper layers than shallow-rooted crops such as corn and sorghum. Weed interference and competition become more severe under moisture stress conditions. Weeds competing with crops under drought conditions contribute to increased crop water stress (Zimdahl 1980; Patterson 1995). Stuart et al. (1984) reported that smooth pigweed (*Amaranthus hybridus* L.) competition with cotton under moisture stress conditions further reduced leaf water potential and turgor pressure in cotton. Varying soil moisture conditions can greatly alter critical periods of weed control (Hartzler 2003). Coble et al. (1981) found that the critical period for the control of common ragweed (*Ambrosia artemisiifolia*) in soybean was only 2 weeks in a drought year, compared to 4 weeks in a wet year. Additionally, certain weed species such as *Echinochloa* spp. and *Cyperus rotundus* show high tolerance to flooding and can have competitive advantage over crops under these conditions (Ismail et al. 2012; Estioko et al. 2014).

The activation, solubility, availability, uptake, metabolism, and efficacy of soil applied herbicides are highly influenced by soil moisture conditions (Muzik 1976; Dickson et al. 1990). Extreme soil moisture conditions can influence crop injury caused by herbicides and ultimately impact crop growth, development, and yields. Inadequate soil moisture following soil-active residual herbicide applications can lead to poor weed control and high weed-crop interference. Drought conditions after application of a soil-active herbicide will restrict herbicide availability to a very narrow band on the soil surface, and weed seedlings emerging through the narrow herbicide band are less likely to be affected (Congreve and Cameron 2015). Drought stress conditions also reduce the uptake and translocation of foliar-applied herbicides. Prolonged drought stress can cause stomata to close, increase cuticle thickness, and/or increase leaf pubescence, thereby reduce herbicide entry into the leaf (Emilio 1991). Steptoe et al. (2006) reported that cuticle thickness and trichome number of Bengal dayflower increased with reduced soil moisture. Likewise, Dickson et al. (1990) found that when oats (*Avena sativa*) were under severe water stress before and after application of fluazifop or glyphosate, no herbicidal injury was observed within 1 month after application (Dickson et al. 1990). While drought conditions do not increase injuries from soil-applied herbicides to the current-season crop, they may reduce herbicide dissipation and increase opportunities for carry-over into subsequent crops; injury may occur if a sensitive crop is planted in rotation. Research by Jebellie et al. (1996) showed that degradation of atrazine was very slow under dry soil conditions, whereas the half-life was only 1 week at 50% soil moisture. Similar findings were reported on the degradation of other herbicides such as hexazinone and simazine (Garcia-Valcarcel and Tadeo 1999).

An interaction of excessive moisture/flooding conditions with herbicide application may also affect herbicide activity on crops and weeds and alter crop-weed interactions and thereby crop yields. Prolonged wet soil conditions can slow down crop growth, leading to herbicide-related crop injuries (Sanchez and Lamont 2012; Hager and Sprague 2002). Crop injuries also occur when high moisture/flooding conditions move soil-applied herbicides to the root zone of the crops (Hartzler 2000) and/or increase the availability of the herbicides in soil water (Hager and Sprague 2002). Excessive moisture/flooding conditions may not only increase crop injury due to direct herbicide uptake by crop roots but also affect crop performance through increased weed growth and interference. Flooding conditions can mobilize soil-applied herbicides below the topsoil layer where the majority of weed seedlings emerge, thereby reducing weed control activity and increasing weed escapes (Majek 2013; Congreve and Cameron 2015).

3.2.3 Impact of Combined Weed/Herbicide and Disease Stress on Crops

Herbicide stress can make the crops vulnerable to diseases, whereas certain herbicide applications can suppress plant pathogens (Altman and Campbell 1977; Levesque and Rahe 1992; Duke et al. 2007; Sanyal and Shrestha 2008). Altman and Ross (1967) suggested that root rot (*Rhizoctonia* spp.) infection was responsible for unexpected preplant herbicide damage on sugar beets. Heydari and Misaghi (1998) studied the impact of PRE application of pendimethalin, trifluralin, and prometryn on *R. solani* and found that the latter herbicide increased the presence of damping-off disease in cotton both under controlled conditions and field situations. Reduction in soybean yield was also observed when metribuzin was applied PRE in *R. solani*-infested soil (Wiley and Ross 1974). Smiley et al. (1992) reported that *R. solani* infection in barley could increase when glyphosate is applied during early crop establishment, leading to yield reduction of up to 50%.

Conversely, herbicide applications may reduce pathogen infection or suppress disease severity. Herbicide applications can sometimes predispose the crops and protect them from subsequent pathogen infections (Levesque and Rahe 1992). Anderson and Kolmer (2005) found that the application of glyphosate may induce a systemic disease resistance response to rust in glyphosate-tolerant wheat. Herbicides can induce the production of potential antibiotic substances or exhibit activity that could weaken at least a certain phase of the pathogen's life cycle (Altman and Campbell 1977). Dann et al. (1999) reported that the herbicide lactofen suppressed *Sclerotinia* stem rot of soybean up to 60%. Similarly, application of glyphosate suppressed rust disease (*Uromyces striatus*) in alfalfa (Samac 2012). Herbicides could also alter a pathogen's biological activity. An example of altered biological activity on pathogens is the exposure of grapevine downy mildew (*Plasmopara viticola*) to low doses of glufosinate. In this case, glufosinate reduces the production of sporangiophores of the downy mildew pathogen by altering the free available amino acids necessary for spore production (Kortekamp 2008). Likewise, reductions in the pathogenicity of Asian soybean rust (*Phakospora pachyrhizi*) was observed when glyphosate was applied on glyphosate-resistant soybeans (Feng et al. 2008). Thus, understanding the combined effect of herbicides and plant pathogens on crops could benefit growers in developing robust management programs for improving crop yields.

3.2.4 Impact of Combined Weed/Herbicide and Insect Stress on Crops

Weeds and insects are both biological constraints and can cause significant crop yield loss. A combined effect of weeds and insects can reduce crop yields either directly through a hosting and feeding relationship or indirectly through insect pests causing damage to crops, thereby reducing crop vigor and diminished competitive ability with weeds. These combined stresses can alter the nature of weed-crop interactions (Norris and Kogan 2000).

Weeds generally act as a colonizer for insects and provide them with resources and shelter during off-season (Norris and Kogan 2000). More than 70 families of potential crop insect pests are associated with weeds (Altieri 1994). Notably, insects such as aphids (Eastop 1981), whiteflies (Yassin and Bendixen 1982), *Heliothis* spp. (Kogan et al. 1989), and lygus bugs (Young 1986) use weeds as alternative hosts for off-season survival. The presence of weed hosts within a crop field can also increase insect damage on the crop. Smitley (1996) reported, for example, that the presence of weeds increased the population of Japanese beetles by tenfold in nursery fields.

Herbicides are often used as an effective management tool to control weeds; however, the presence of certain insects may alter crop-weed interaction with herbicides. A study showed that stalk-boring insects enhanced glyphosate activity on 15-cm tall giant ragweed (*Ambrosia trifida*) plants, whereas decreased the efficacy of glyphosate on 45-cm tall plants (Ott et al. 2007). Westra and Wyse (1978) reported that the efficacy of glyphosate on quack grass (*Agropyron repens*) was reduced when infested by the weevil *Notaris bimaculatus*. Similarly, Harder et al. (2007) found that glyphosate application combined with the presence of leaf miner and beet petiole borer (*Cosmobaris americana*) drastically reduced the control of common lambsquarters (*Chenopodium album*). On the other hand, combination of insects and herbicides has been shown to improve weed control. Boydston and Williams (2004) showed that the combination of gall mite (*Aceria malherbae*) and herbicides (2,4-D or glyphosate) can greatly reduce the root biomass of field bindweed (*Convolvulus arvensis*), compared to herbicide alone.

Herbicide application can also have a negative impact on insects. Application of chlorsulfuron to wild buckwheat (*Polygonum convolvulus*) reduced the fitness of the leaf-eating beetle *Gastrophysa polygoni* (Kjaer and Elmegaard 1996). Alachlor can provide partial control of corn rootworms (*Diabrotica* spp.) (Gray et al. 1984; Reed et al. 1989), whereas ingestion of asulam-contaminated leaves reduced the fecundity of female *Gastrophysa viridula* beetles by 64% (Speight and Whittaker 1987).

3.3 Conclusion

This review establishes that combination of weed/herbicide stress with other biotic and abiotic stressors such as temperature, moisture, disease, and insects can greatly alter the overall impact and outcome on crop yields. There are few other stress factors such as salinity (Islam et al. 2016) that may interact with weeds/herbicides and influence crop response, but are not reviewed here. In many cases, the combination of weed/herbicide stress and other stressors has been shown to accelerate crop yield loss. Yet, little is understood on the nature and dynamics of such interactions. More research is necessary in this arena to fully understand multiple stress interactions involving weeds/herbicides and utilize that knowledge for improving crop productivity.

References

Alberto AMP, Ziska LH, Cervanica CR, Manalo PA (1996) The influence of increasing carbon dioxide and temperature on competitive interactions between a C_3 crop, Rice and a C_4 weed (*Echinochloa glabrescens*). Aust J Plant Physiol 23:795–802

Allen DJ, Ort DR (2001) Impacts of chilling temperatures on photosynthesis in warm-climate plants. Trends Plant Sci 6:36–42

Altieri MA (1994) Biodiversity and pest management. Food Products Press, New York, p 195

Altman J, Campbell CL (1977) Effect of herbicides on plant diseases. Annu Rev Phytopathol 15:361–365

Altman J, Ross M (1967) Plant pathogens as possible factor in unexpected preplant herbicide damage in sugarbeets. Plant Dis Rep 51:86–88

Anderson JA, Kolmer JA (2005) Rust control in glyphosate tolerant wheat following application of the herbicide glyphosate. Plant Dis 89:1136–1142

Barlow K, Christy B, O'leary G, Riffkin P, Nuttall J (2015) Simulating the impact of extreme heat and frost events on wheat crop production: a review. Field Crop Res 171:109–119

Barrentine WL (1974) Common cocklebur competition in soybeans. Weed Sci 22:600–603

Bloomberg JR, Kirkpatrick BL, Wax LM (1982) Competition of common cocklebur (*Xanthium pensylvanicum*) with soybean (*Glycine max*). Weed Sci 30:507–513

Blum A (2011) Plant breeding for water limited environments. Springer XII p 255

Boydston RA, Williams MM (2004) Combined effects of Acedia malherbae and herbicides on field bindweed (*Convolvulus arvensis*) growth. Weed Sci 52:297–301

Bradley CA, Hartman GL, Wax LM, Pedersen WL (2002) Influence of herbicides on Rhizoctonia root and hypocotyl rot of soybean. Crop Prot 21:679–687

Coble HD, Williams FM, Ritter RL (1981) Common ragweed (*Ambrosia artemisiifolia*) interference in soybeans (*Glycine max*). Weed Sci 29:339–342

Congreve M, Cameron J (2015) Soil behavior of pre-emergence herbicides in Australian farming systems- a reference manual for agronomic advisers. https://grdc.com.au. Accessed 11 Aug 2016

Dann EK, Diers BW, Hammerschmidt R (1999) Suppression of Sclerotinia stem rot of soybean by lactofen herbicide treatment. Phytopathology 89:598–692

Dickson RL, Andrews M, Field RJ, Dickson EL (1990) Effect of water stress, nitrogen, and gibberellic acid on fluazifop and glyphosate activity on oats (*Avena sativa*). Weed Sci 38:54–61

Duke SO, Wedge DE, Cerdeira AL, Matallo MB (2007) Herbicide effects on plant disease. Outlooks Pest Manag 18:36–40

Eastop VF (1981) The wild hosts of aphid pests. In: Thresh JM (ed) Pests, pathogens and vegetation. Pitman Books Limited, London, pp 285–298

Emilio SO (1991) Effect of weed water stress on postemergence herbicides activity. Dissertation, Iowa State University

Estioko LP, Miro B, Baltazar AM, Merca FE, Ismail AM, Johnson DE (2014) Differences in responses to flooding by germinating seeds of two contrasting rice cultivars and two species of economically important grass weeds. AoB Plants:1–15

Evans LT (1993) Crop evolution, adaptation and yield. Cambridge University Press, Cambridge

Fausey JC, Renner KA (2001) Environmental effects on CGA-248757 and flumiclorac efficacy/soybean tolerance. Weed Sci 49:668–674

Feng PCC, Clark C, Andrade GC, Balbi MC, Caldwell P (2008) The control of Asian rust by glyphosate in glyphosate-resistant soybeans. Pest Manag Sci 64:353–359

Flint EP, Patterson DT (1983) Interference and temperature effects on growth in soybean (*Glycine max*) and associated C_3 and C_4 weeds. Weed Sci 1:193–199

Flint EP, Patterson DT, Beyers JL (1983) Interference and temperature effects on growth of cotton (*Gossypium hirsutum*), spurred anoda (*Anoda cristata*), and velvetleaf (*Abutilon theophrasti*). Weed Sci 31:892–898

Garcia-Valcarcel AI, Tadeo JL (1999) Influence of soil moisture on sorption and degradation of hexazinone and simazine in soil. J Agric Food Chem 47:3895–3900

Ge X, d'Avignon DA, Ackerman JJ, Duncan B, Spaur MB, Sammons RD (2011) Glyphosate-resistant horseweed made sensitive to glyphosate: low-temperature suppression of glyphosate vacuolar sequestration revealed by 31P NMR. Pest Manag Sci 67:1215–1221

Godar AS, Varanasi VK, Nakka S, Prasad PVV, Thompson CR, Mithila J (2015) Physiological and molecular mechanisms of differential sensitivity of Palmer amaranth (*Amaranthus palmeri*) to mesotrione at varying growth temperatures. PLoS One 10:e0126731. doi:10.1371/journal.pone.0126731

Gray ME, Coats JR, Tollefsen JJ (1984) Effect of an insecticide and a herbicide combination on corn rootworm (Coleoptera: Chrysomelidae) damage. J Econ Entomol 77:465–467

Hager A, Sprague C (2002) Factors contributing to the likelihood of corn injury. http://bulletin.ipm.illinois.edu. Accessed 11 Aug 2016

Hall MR, Swanton CJ, Anderson GW (1992) The critical period of weed control in grain corn (*Zea mays*). Weed Sci 40:441–447

Harder DB, Sprague CL, Difonzo CD, Renner KA, Ott EJ, Johnson WG (2007) Influence of stem-boring insects on common lambsquarters (*Chenopodium album*) control in soybean with glyphosate. Weed Technol 21:241–248

Harrison HF, Peterson JK (1999) Effect of temperature and cultivar on the response of broccoli and collard (*Brassica oleracea*) to oxyfluorfen. Weed Technol 13:726–730

Hartzler B (2000) Crop response to herbicides. http://www.weeds.iastate.edu. Accessed 11 Aug 2016

Hartzler B (2003) Critical periods of competition. http://www.weeds.iastate.edu. Accessed 28 Sept 2016

Hatfield JL, Prueger JH (2015) Temperature extremes: effect on plant growth and development. Weather Clim Extremes 10:4–10. doi:10.1016/j.wace.2015.08.001

Heydari A, Misaghi IJ (1998) The impact of herbicides on the incidence and development of Rhizoctonia solani induced cotton seedling damping-off. Plant Dis 82:110–113

Islam F, Ali B, Wang J, Farooq MA, Gill RA, Ali S, Wang D, Zhow W (2016) Combined herbicide and saline stress differentially modulates hormonal regulation and antioxidant defense system in *Oryza sativa* cultivars. Plant Physiol Biochem 107:82–95

Ismail AM, Johnson DE, Ella ES, Vergara GV, Baltazar AM (2012) Adaptation of flooding during emergence and seedling growth in rice and weeds, and implications for crop establishment. AoB Plants:1–18

Jaleel CA, Manivannan P, Wahid A, Farooq M, Al-Juburi HJ, Somasundaram R, Panneerselvam R (2009) Drought stress in plants: a review on morphological characters and pigments composition. Int J Agric Biol 11:100–105

Jebellie SJ, Prasher SO, Clemente RS (1996) Role of soil moisture content in reducing environmental pollution from pesticides. Can Water Resour J 21:393–402

Jia Q, Lv B, Guo M, Luo C, Zheng L, Hsiang T, Huang J (2015) Effect of rice growth stage, temperature, relative humidity and wetness duration on infection of rice panicles by Villosiclava virens. Eur J Plant Pathol 141:15–25

Jordan TN (1977) Effects of temperature and relative humidity on the toxicity of glyphosate to bermudagrass (*Cynodon dactylon*). Weed Sci 25:448–451

Kamara AY, Menkir A, Badu–Apraku B, Ibikunle O (2003) The influence of drought stress on growth, yield and yield components of selected maize genotypes. J Agric Sci 141:43–50

Kirnak H, Kaya C, TAS I, Higgs D (2001) The influence of water deficit on vegetative growth, physiology, fruit yield and quality in eggplants. Bulg J Plant Physiol 27:34–46

Kjaer C, Elmegaard N (1996) Effect of herbicide treatment on host plant quality for a leaf-eating beetle. Pestic Sci 47:319–325

Kogan MC, Helm CG, Kogan J, Brewer E (1989) Distribution and economic importance of Heliothis virescens and Heliothis sea in North Central, and South America and of their natural enemies and host plants. Proceedings of the workshop on Biological Control of Heliothis: Increasing the Effectiveness of Natural Enemies, New Delhi, India. Office of International Cooperation & Development, USDA, Washington DC, pp 241–297

Kortekamp A (2008) Knocked out with Basta!-Are herbicides effective against downy mildew of grapevine? J Plant Dis Protect Special Issue XXI 107–112. ISSN 1861–4051

Kramer PJ (1980) Drought, stress, and the origin of adaptations. In: Turner NC, Kramer PJ (eds) Adaptation of plants to water and high temperature stress. Wiley, New York, pp 7–20

Lee SD, Oliver LR (1982) Efficacy of acifluorfen on broadleaf weeds. Times and methods of application. Weed Sci 30:520–526

Legg BJ (1983) Micrometeorology and the influence of local variations of environment on plant growth and herbicide performance. Asp Appl Biol 4:15–31

Levesque CA, Rahe JE (1992) Herbicide interactions with fungal root pathogens, with special reference to glyphosate. Annu Rev Phytopathol 30:579–602

Majek B (2013) How rainfall influences residual herbicides. http://plant-pest.advisory.rutgers.edu. Accessed 11 Aug 2016

Manikavelu A, Nadarajan N, Ganesh SK, Gnanamalar RP, Babu RC (2006) Drought tolerance in rice: morphological and molecular genetic consideration. Plant Growth Regul 50:121–138

Masiunas JB, Weller SC (1988) Glyphosate activity in potato (*Solanum tuberosum*) under different temperature regimes and light levels. Weed Sci 1:137–140

Matzenbacher FO, Vidal RA, Merotto A, Trezzi MM (2014) Environmental and physiological factors that affect the efficacy of herbicides that inhibit the enzyme protoporphyrinogen oxidase: a literature review. Planta Dan 32:457–463

Mazahery-Laghab H, Nouri F, Abianeh HZ (2003) Effects of the reduction of drought stress using supplementary irrigation for sunflower (*Helianthus annuus*) in dry farming conditions, Pajouheshva- Sazandegi. Agric Hortic 59:81–86

Mesihovic A, Iannacone R, Firon N, Fragkostefanakis S (2016) Heat stress regimes for the investigation of pollen thermotolerance in crop plants. Plant Reprod 25:1–13

Monneveux P, Sanchez C, Beck D, Edmeades GO (2006) Drought tolerance improvement in tropical maize source populations: evidence of progress. Crop Sci 46:180–191

Mundt CC, Brophy LS, Schmitt MS (1995) Disease severity and yield of pure-line wheat cultivars and mixtures in the presence of eyespot, yellow rust, and their combination. Plant Pathol 44:173–182

Murphy SD, Yakubu Y, Weise SF, Swanton CJ (1996) Effect of planting patterns and inter-row cultivation on competition between corn (*Zea mays*) and late emerging weeds. Weed Sci 1:865–870

Muzik TJ (1976) Influence of environmental factors on toxicity to plants. In: Audus LJ (ed) Herbicides. Physiology, biochemistry, ecology. Academic Press, New York, pp 203–247

Nayyar H, Kaur S, Singh S, Upadhyaya HD (2006) Differential sensitivity of Desi (small-seeded) and Kabuli (large-seeded) chickpea genotypes to water stress during seed filling: effects on accumulation of seed reserves and yield. J Sci Food Agric 86:2076–2082

Nie GY, Long SP, Baker NR (1992) The effects of development at sub-optimal growth temperatures on photosynthetic capacity and susceptibility to chilling-dependent photoinhibition in *Zea mays*. Physiol Plant 85:554–560

Norris RF, Kogan M (2000) Interactions between weeds, arthropod pests, and their natural enemies in managed ecosystems. Weed Sci 48:94–158

Obilo OP, Ikotun B, Ihejirika GO, Ibeawuchi II, Oben TT (2010) The effect of the incidence of cassava anthracnose disease (CAD) on the performance and yield of cassava cultivars. Crop Prot 29:482–486

Oerke EC (2006) Crop losses to pests. J Agric Sci 144:31–43

Ott EJ, Gerber CK, Harder DB, Sprague CL, Johnson WG (2007) Prevalence and influence of stalk-boring insects on glyphosate activity on Indiana and Michigan giant ragweed (*Ambrosia trifida*). Weed Technol 21:526–531

Ozturk M, Rehder H, Zeigler H (1981) Biomass production of C_3 and C_4 plant species in pure and mixed culture with different water supply. Oecologia 50:73–81

Parminder SC, Jatinder SA, Jugulam M, Amit JJ (2015) Herbicide-resistant Palmer amaranth (*Amaranthus palmeri* S. Wats.) in the United States- mechanisms of resistance, impact, and management. In: Andrew Price, Jessica Kelton and Lina Sarunaite (ed) Herbicides, Agronomic Crops and Weed Biology. Intech Publisher, pp 1–29

Patterson DT (1995) Effects of environmental stress on weed/crop interactions. Weed Sci 43:483–490

Peng S, Huang J, Sheehy JE, Laza RC, Visperas RM, Zhong X, Centeno GS, Khush GS, Cassman KG (2004) Rice yields decline with higher night temperature from global warming. PNAS 101:9971–9975

Pillmoor JB (1985) Influence of temperature on the activity of AC 222, 293 against *Avena fatua* L. and *Alopecurus myosuroides* Huds. Weed Res 25:433–442

Porter JR, Semenov MA (2005) Crop responses to climatic variation. Philos Trans R Soc Lond B Biol Sci 360:2021–2035

Potter JR, Jones JW (1977) Leaf area partitioning as an important factor in growth. Plant Physiol 59:10–14

Prasad PV, Craufurd PQ, Summerfield RJ, Wheeler TR (2000) Effects of short episodes of heat stress on flower production and fruit-set of groundnut (*Arachis hypogaea* L.). J Exp Bot 51:777–784

Prasad PV, Boote KJ, Allen LH (2006a) Adverse high temperature effects on pollen viability, seed-set, seed yield and harvest index of grain-sorghum [*Sorghum bicolor* (L.) Moench] are more severe at elevated carbon dioxide due to higher tissue temperatures. Agric For Meteorol 139:237–251

Prasad PV, Boote KJ, Allen LH, Sheehy JE, Thomas JM (2006b) Species, ecotype and cultivar differences in spikelet fertility and harvest index of rice in response to high temperature stress. Field Crop Res 95:398–411

Price CE (1983) The effect of environment on foliage uptake and translocation of herbicides. Asp Appl Biol 4:157–169

Reed JP, Keaster AJ, Kremer RJ, Krause GF (1989) Synergistic and antagonistic responses of soil insecticide-herbicide combinations for corn rootworm Diabrotica spp. control. J Environ Health Sci B 24:325–334

Roberts EH, Qi A, Ellis RH, Summerfield RJ, Lawn RJ, Shanmugasundaram S (1996) Use of field observation to characterise genotypic flowering responses to photoperiod and temperature: a soyabean exemplar. Theor Appl Genet 93:519–533

Rymen B, Fiorani F, Kartal F, Vandepoele K, Inzé D, Beemster GT (2007) Cold nights impair leaf growth and cell cycle progression in maize through transcriptional changes of cell cycle genes. Plant Physiol 143:1429–1438

Samac DA (2012) Effect of glyphosate application on foliar diseases in glyphosate-tolerant alfalfa. Plant Dis 96:1104–1110

Samarah NH, Mullen RE, Cianzio SR, Scott P (2006) Dehydrin-like proteins in soybean seeds in response to drought stress during seed filling. Crop Sci 46:2141–2150

Sanchez E and Lamont B (2012) Herbicide injury as a result of cool, wet weather conditions. http://extension.psu.edu. Accessed 8 Oct 2016

Sanyal D, Shrestha A (2008) Direct effect of herbicides on plant pathogens and disease development in various cropping systems. Weed Sci 56:155–160

Seiter NJ, Greene JK, Reay-jones FPF (2013) Reduction of soybean yield components by Megacopta cribraria (Hemiptera: Plataspidae). J Econ Entomol 106:1676–1683

Smiley RW, Ogg AG Jr, Cook RJ (1992) Influence of glyphosate on Rhizoctonia root rot, growth, and yield of barley. Plant Dis 76:937–942

Smitley DR (1996) Incidence of *Popillia japonica* (Coleoptera: Scarabaeidae) and other scarab larvae in nursery fields. J Econ Entomol 89:1262–1266

Speight RI, Whittaker JB (1987) Interactions between the Chrysomelid beetle Gastrophysa viridula, the weed *Rumex obtusifolius*, and the herbicide asulam. J Appl Ecol 24:119–129

Steduto P, Hsiao TC, Fereres E, Raes D (2012) Crop yield response to water. http://www.fao.org/nr/water/docs/irrigationdrainage66.pdf. Accessed 11 Aug 2016

Steptoe PJ, Vencill WK, Grey TL (2006) Influence of moisture stress on herbicidal control of an invasive weed, Benghal dayflower (Commelina bhengalensis). J Plant Dis Protect 20:907–914

Stuart BL, Harrison SK, Abernathy JR, Krieg DR, Wendt CW (1984) The response of cotton (*Gossypium hirsutum* L.) water relations to smooth pigweed (*Amaranthus hybridus*) competition. Weed Sci 32:126–132

Teixeira EI, Fischer G, van Velthuizen H, Walter C, Ewert F (2013) Global hot-spots of heat stress on agricultural crops due to climate change. Agric For Meteorol 170:206–215

Thompson L Jr, Slife FW, Butler HS (1970) Environmental influence on the tolerance of corn to atrazine. Weed Sci 1:509–514

Tiaz L, Zeiger E (1991) Plant physiology. The Benjamin/Cummings Publishing Company, New York, pp 219–248

Viger PR, Eberlein CV, Fuerst EP (1991) Influence of available soil water content, temperature, and CGA-154281 on metolachlor injury to corn. Weed Sci 39:227–231

Westra P, Wyse DL (1978) Control of quackgrass biotypes with glyphosate. Proc North Cent Weed Control Conf 33:106

Wheeler TR, Craufurd PQ, Ellis RH, Porter JR, Prasad PV (2000) Temperature variability and the yield of annual crops. Agric Ecosyst Environ 82:159–167

Wiley GL, Ross MA (1974) Effect of herbicides on Rhizoctonia root rot of soybeans. 29th Ann Res Rept North Cent Weed Cont Conf, pp 33–34

Xie HS, Hsiao AI, Quick WA (1996) Influence of temperature and light intensity on absorption, translocation, and phytotoxicity of fenoxaprop-ethyl and imazamethabenz-methyl in *Avena fatua*. J Plant Growth Regul 15:57–62

Yamori W, Hikosaka K, Way DA (2014) Temperature response of photosynthesis in C_3, C_4, and CAM plants: temperature acclimation and temperature adaptation. Photosynth Res 119:101–117

Yassin M, Bendixen LE (1982) Weed hosts of the cotton whitefly (*Bemisia tabaci* (Genn.)) Homoptera: Aleyrodidae. Wooster, OH: The Ohio State University, Ohio Agricultural Research and Development Center Research Bulletin 1144

Young OP (1986) Host plants of the tarnished plant bug, *Lygus lineolaris* (Heteroptera: Miridae). Ann Entomol Soc Am 79:747–762

Zimdahl RL (1980) Weed-crop competition. A review. International Plant Protection Center. Oregon State University, Corvallis, p 195

Heat and Soil Moisture Stress Differentially Impact Chickpea Plant Infection with Fungal Pathogens

4

Mamta Sharma and Raju Ghosh

Abstract

Plants are often simultaneously exposed to multiple biotic and abiotic stresses resulting in substantial yield loss. Moreover, increase in the frequency of climate extremes is likely to influence the distribution, establishment and epidemiology of plant diseases. Emerging evidences suggest the changing scenario of diseases in chickpea, a grain legume largely grown in rain-fed environments. In this chapter, we have focused on the major and emerging soil-borne diseases in chickpea that are largely influenced by differential temperature and soil moisture stress. Changes in the disease spectrum in chickpea for the past one decade were monitored through extensive surveys. Analysis of disease and weather data indicated shift in the occurrence and distribution of chickpea diseases as well as emergence of new diseases. Dry root rot (*Rhizoctonia bataticola*) is becoming more intense in tropical humid areas under high temperature and soil moisture stress. Contrary to this, sporadic occurrence of collar rot (*Sclerotium rolfsii*) has been noticed under high soil moisture levels. Host resistance influenced by soil moisture levels and rise in temperature have also been discussed. Extensive research is required in this domain to develop adaptation and mitigation strategies for sustained food security. Breeding being an essential part of crop improvement needs to keep pace with these emerging diseases.

Keywords

Biotic stress • Chickpea • Climate change • Host resistance • Soil moisture • Temperature

M. Sharma (✉) • R. Ghosh
International Crops Research Institute for the Semi-Arid Tropics (ICRISAT), Patancheru, Hyderabad 502 324, Telangana, India
e-mail: mamta.sharma@cgiar.org

M. Senthil-Kumar (eds.), *Plant Tolerance to Individual and Concurrent Stresses*,
DOI 10.1007/978-81-322-3706-8_4

4.1 Introduction

Chickpea (*Cicer arietinum* L.) is one of the essential semiarid tropical legume crops and is either grown in the post-rainy season on stored soil moisture (south Asia and spring-sown Mediterranean) or as a Mediterranean winter crop. In both of these instances, the crop is exposed to terminal drought accompanied with high temperatures. Under the changing scenario of climate, more erratic rainfall patterns and spells of temperature extremes will consequently affect the crop productivity (Graham and Vance 2003). A steady increase in temperature, decrease/increase in relative humidity and moisture stress will not only affect the crop *per se* but will also change the relative activity and abundance of diseases, natural enemies, and their interaction with the host plants. As a consequence of it, shift in the chickpea diseases have been seen in the major growing regions worldwide.

Chickpea is a short-duration, self-pollinated, diploid ($2n = 2x = 16$) legume with genome size of 738 Mb (Varshney et al. 2013). It is cultivated in different parts of the world mainly in the Mediterranean, South Asia, North Africa, Middle East and North and Central America. It is a rich and cheap source of vegetarian protein, vitamins and minerals (Jukanti et al. 2012), making it nutritionally more valuable. These valuable aspects of chickpea caused an increase in its global cultivation and the overall production reached 10.0 million metric tons (mt) from 6.6 mt (http://www.cgiar.org/ourresearch/crop-factsheets/chickpea, as on 23rd April 2013) during the last 30 years. South Asia accounts for more than 75% of the total area under chickpea cultivation, and India is the world leader in chickpea production with 7.5 mt (FAOSTAT 2010), followed by Pakistan and Turkey.

Both biotic and abiotic stresses cause significant yield losses (corresponding to 11.2 mt) in chickpea (Ryan 1997). Among the abiotic stresses, drought causes a 40–50% reduction in yield globally (Graham and Vance 2003) and is emerging as a major barrier to its wider cultivation on the drought prone semiarid tropic (SAT) region. With the increasing drought, fungal diseases like dry root rot that thrive in the drought conditions are emerging as major threat to its production. Temperature may have important repercussions on the effectiveness of host plant resistance. Theoretically, following three types of abiotic–biotic interactions can be expected in chickpea:

- A direct effect on pathogens
- Indirect effect on pathogens through other community interactions
- Interaction effects through host physiology, i.e. multiple stress concept

This chapter deals with the soil-borne diseases and pathogens of chickpea being largely impacted by temperature and soil moisture stresses within the production situations. We have highlighted sequential occurrence of chickpea diseases under different weather scenarios in SAT regions, followed by short narratives of emerging diseases and few evidences/data that support these results and finally conclusions and presumption for the future.

4.2 Sequential Occurrence of Soil-Borne Diseases of Chickpea in SAT Environments

The spatial and temporal succession of soil-borne diseases such as Fusarium wilt (*Fusarium oxysporum* f. sp. *ciceris*), dry root rot (*Rhizoctonia bataticola*), collar rot (*Sclerotium rolfsii*), black root rot (*Fusarium solani*) and wet root rot (*Rhizoctonia solani*) of chickpea in SAT is strongly influenced by the prevailing climate variables. An analysis of the weather data and diseases pattern in chickpea in the past one decade indicated a shift in the disease pattern (Sharma 2012). The production of chickpea is largely constrained by Fusarium wilt in crop season with no drastic variations in weather, and, therefore, all the breeding efforts in past were towards developing wilt resistant cultivars. As a result, several wilt resistant cultivars have been deployed and released worldwide (Gowda and Gaur 2004). However, in present scenario frequency of the occurrence of diseases like collar rot and dry root rot has increased due to change in temperature and rainfall. Past and present scenarios of chickpea diseases as influenced by environment are represented in Figs. 4.1 and 4.2. Figure 4.1 signifies the predominance of Fusarium wilt throughout the crop season provided the weather conditions have no drastic variations, and the Fig. 4.2 illustrates the predominance of collar rot at the seedling stage and dry root rot at the time of flowering and podding predisposed by high moisture and drought, respectively. The later scenario has become more common in SAT, thereby demanding the research focus on emerging diseases with respect to their

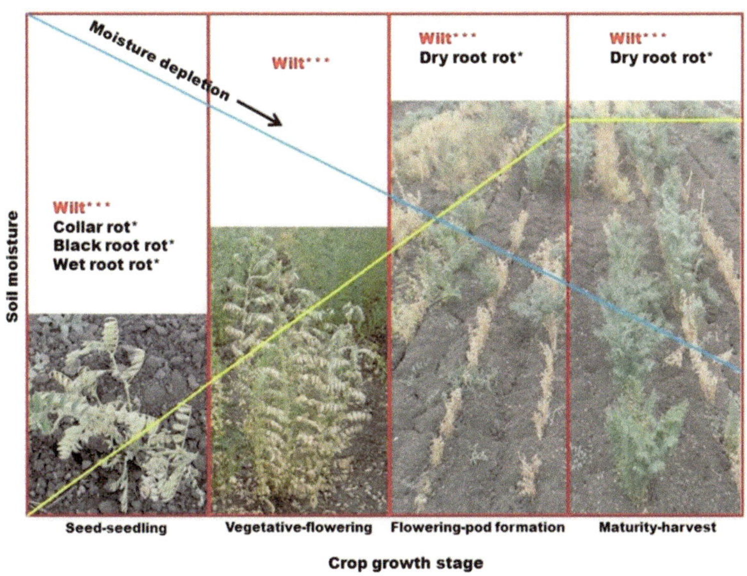

Fig. 4.1 Past scenario of chickpea diseases with respect to past weather (*indicates less incidence; ***indicate more incidence of the disease)

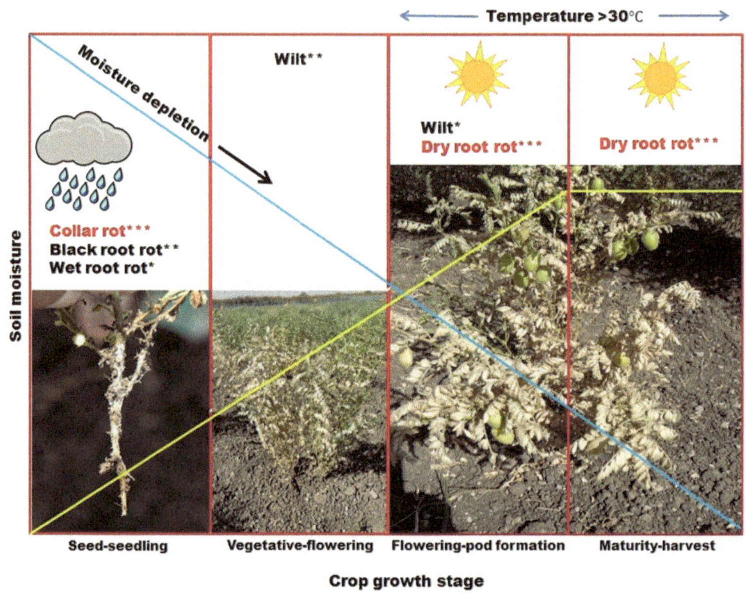

Fig. 4.2 Present scenario of chickpea diseases with respect to changed weather (*indicates less incidence; ***indicate more incidence)

epidemiology, pathogen biology and host × pathogen × environment interactions. In the following section, two examples of chickpea diseases influenced by drought and high soil moisture have been discussed.

4.2.1 High Temperature and Soil Moisture Stress Predisposing Factor for Dry Root Rot

Dry root rot caused *by Rhizoctonia bataticola* (*Macrophomina phaseolina*) is an important component of the disease complex that causes root rots and seedling blight in many grain legumes when they are weakened by other stress factors (Hwang et al. 2003). *R. bataticola* is a soil inhabiting organism and most commonly infects chickpea at post-reproductive stage in dry and warm regions (Sharma and Pande 2013). The disease is of increasing importance in chickpea and has the potential to cause devastation in susceptible cultivars, particularly in the conditions of high temperature and soil moisture stress. Savary et al. (2011) described dry root rot as an acute-emerging disease that occurs irregularly, both temporally and spatially, and may cause massive disruptions in system performances and whose range is expanding to new areas.

The disease has been reported from most chickpea growing areas in India and other countries like Iran, the USA and several countries in Asia and Africa (see details in Sharma et al. 2015) but has become a major threat to chickpea production in recent years due to longer drought spells at the time of flowering and podding (Sharma and Pande 2013). Ghosh et al. (2013) conducted a survey during 2010–2013 in India and indicated widespread and increased incidence of dry root rot in the central and southern states of India. The disease was found irrespective of soil types, cropping system and cultivars used and incidence ranged from 5 to 50% or more in badly infected soils. This noticeable widespread geographic distribution of dry root rot probably makes it a significant disease in chickpea. *R. bataticola* is more virulent under high temperatures (32 °C) and as a result cause severe damage on chickpeas in the warmer Salinas Valley in California (Raabe 1985).

Sharma and Pande (2013) demonstrated the relationship between temperature and soil moisture stress on the development of dry root rot. They conducted series of experiments in controlled environment conditions to understand the effect of temperature and soil moisture alone or in combination on infection, colonisation and development of dry root rot. They concluded that a combination of high temperature (≥30 °C) and soil moisture content (≤60%) are positively correlated with dry root rot incidence/severity in chickpea (Fig. 4.3). Singh and Sharma (2002) also reported that soil moisture deficit favours the severe disease development on pulse crops. Under hot and dry conditions, many economically important crops are predisposed to *R. bataticola* infection such as soybean (Pearson et al. 1984), sunflower (Nawaz

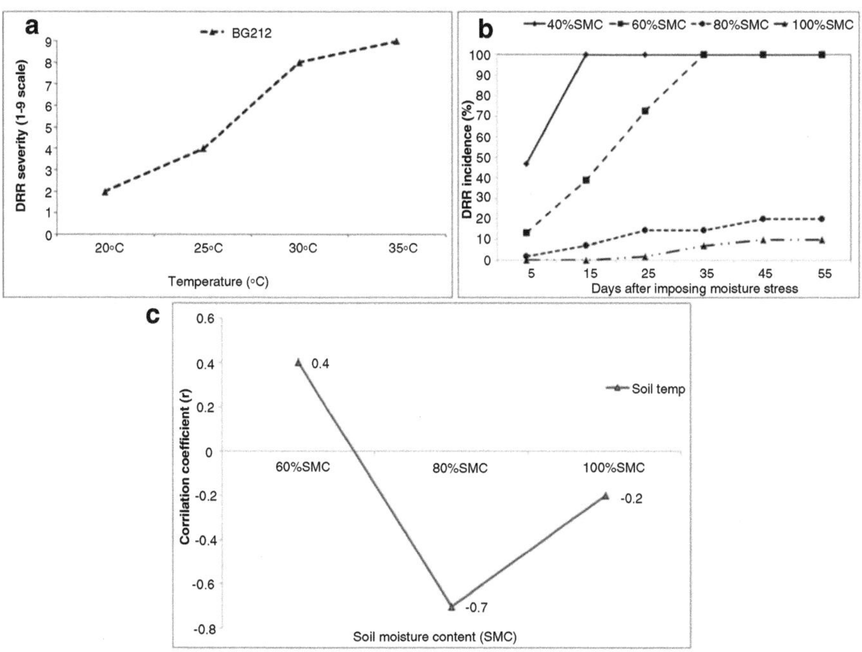

Fig. 4.3 Effect of temperature and soil moisture (**a**) on dry root rot severity (1–9 scale) (**b**) and on incidence of dry root rot (**c**) correlation between soil moisture content, soil temperature and dry root rot incidence

Khan 2007), etc. Sclerotial development was affected by soil moisture levels; more sclerotia were found under relatively low water potentials favouring the disease development (Olaya and Abawi 1996); however the effect was adverse under high water potential.

4.2.2 High Soil Moisture Predisposing Factor for Collar Rot

Collar rot (*S. rolfsii)* is a soil-borne disease, emerging as a potential threat to chickpea production in the tropics, subtropics and other warm temperate regions. The disease in chickpea is favoured by abundant soil moisture, high soil temperature (25–30 °C) and low organic matter in the soil and has a major impact on plant fitness/growth. The disease can cause 55–95% mortality of chickpea seedlings under favourable environmental conditions. Pathogen's extensive host range (at least 500 species in 100 families) most commonly in legumes, crucifers and cucurbits further makes this disease challenging. Ghosh et al. (2013) reported collar rot as an emerging disease in chickpea particularly in years when there is heavy rainfall at the seedling stage, leading to high moisture conducive for the disease development.

S. rolfsii grows, survives and attacks plants at or near the soil line. The fungus can overwinter as mycelium in infected tissues or plant debris. Sclerotia serve as the principal overwintering structure and primary inoculum for disease persistence near the soil surface. Sclerotia may exist free in soil or in association with plant debris. Those buried deep in the soil may survive for a year or less, whereas those at surface remain viable and may germinate in response to volatile compounds released from decomposing plant material. Before the pathogen penetrates host tissue, it produces a considerable mass of mycelium on the plant surface, a process which can take 2–10 days. Production of mycelium and the formation of sclerotia rely upon favourable environmental conditions. Hussain et al. (2006) showed that there was a positive correlation between disease severity and inoculum concentrations where seedling mortality increased with an increase in inoculum load. Lack of information about factors that affect the development of collar rot has made control of this disease difficult.

4.3 Host Resistance Influenced by Changes in Temperature and Other Factors

The influence of temperature on expression of resistance or susceptibility to plant pathogens has been shown by various researchers (Ash and Rees 1994; Brake et al. 1995; French and Elder 1996; Ge et al. 1998; Harling et al. 1998; Judelson and Michelmore 1992; Kim and Bockus 2003; Omwega and Roberts 1992; Sydenham et al. 1997). Fusarium wilt (*F. oxysporum* f. sp. *ciceris*) in combination with cyst nematode is already on rise in some countries probably due to soil temperature rise (Jimenez-Díaz et al. 1993). Also, the race-specific resistant response of chickpea cultivars to infection by *F. oxysporum* f. sp. *ciceris* was reported to be significantly influenced by the increase in temperature (Landa et al. 2006). Artificial inoculation experiments showed that a 3 °C increase, from 24 to 27 °C, in the incubation temperature was sufficient for the reaction of kabuli cv. Ayala and accession PV-1 to race 1A to shift from moderately/highly resistant at constant 24 °C to highly susceptible at 27 °C. A similar but less pronounced effect was found for 'Ayala' infected with race 6 (Landa et al. 2006). However, the susceptible reaction of accession JG-62 to races 1A and 6 was not influenced by the temperature increase. This temperature effect has an impact on the use of cultural practices for management of Fusarium wilt of chickpea. High level of resistance of 'Ayala' to Fusarium wilt when sown in mid- to late January differed from a moderately susceptible reaction under warmer temperatures when sowing was delayed to late February or early March (Landa et al. 2006) indicating that resistance in this cultivar may be temperature dependent and that warmer temperatures, associated with later sowings, may affect the disease reaction of this cultivar. This study demonstrates the importance of temperature in identifying resistant genotypes and races of the pathogen, as well as choosing sowing dates and using resistant chickpea genotypes for the management of Fusarium wilt in different growing areas.

Temperature can also influence the plant–rhizobacteria interactions related to biocontrol potential. For instance, in chickpea, seed and soil treatment with *Pseudomonas fluorescens* RGAF 19, *P. fluorescens* RG 26, *Bacillus megaterium* RGAF 51 and *Paenibacillus macerans* RGAF 101 can suppress Fusarium wilt, but the extent of disease suppression by these rhizobacteria is modulated by soil temperature. *Pseudomonas fluorescens* isolates significantly increased chickpea shoot dry weight at 20 °C and root dry weight at 25 and 30 °C (Landa et al. 2004). All bacterial isolates colonised the chickpea rhizosphere and internal stem tissues at 20, 25 and 30 °C, and there was a positive linear trend between bacterial population size in the rhizosphere and temperature increase. The maximum inhibition of mycelial growth and conidial germination of *Fusarium oxysporum* f. sp. *ciceris* race 5 in vitro occurred at a temperature range optimal for bacterial growth and production of inhibitory metabolites. These results demonstrate the need to understand the effects of environmental factors on the biological activities of introduced rhizobacteria of significant importance for plant disease suppression.

4.4 Effect of Drought and Moisture on Plant–Pathogen Interactions at Biochemical and Molecular Level

Plant responses to different stresses are highly complex and involve changes at the cellular, physiological and transcriptome levels. It has been found that plants respond to multiple stresses differently from how they do to individual stresses, activating a specific programme of gene expression relating to the exact environmental conditions encountered. Rather than being additive, the presence of an abiotic stress can have the effect of reducing or enhancing susceptibility to a biotic pest or pathogen, and vice versa. This interaction between biotic and abiotic stresses is orchestrated by hormone signalling pathways that may induce or antagonise one another. Specificity in multiple stress responses is further controlled by a range of molecular mechanisms that act together in a complex regulatory network (Atkinson and Urwin 2012).

The studies on plant responses to abiotic stresses particularly heat and drought at the biochemical and molecular level have advanced considerably in recent years and have shown that abiotic stresses impact responses to pathogens in several crops. The impact of concurrent drought stress and pathogen infection on plants has been recently documented by Pandey et al. (2014). Although no such studies have been reported on pathogens infecting chickpea so far; however it was found in preliminary studies on *Rhizoctonia bataticola*-chickpea pathosystem that antioxidant enzymes like PAL, PPO, POD and phenol increased under moisture stress as compared to high moisture levels in sick soils (Sharma et al. unpublished data).

Genes and signalling pathways involved in resistance to pathogens have been unravelled in some of the model plants (Dangl and Jones 2001, Wan et al. 2002). In *Arabidopsis,* the Early Responsive to Dehydration 15 gene (*ERD15*) is rapidly induced in response to drought and pathogen infection (Kariola et al. 2006). The overexpression and silencing of the *ERD15* gene not only affected abiotic stress tolerance but also disease resistance. Drought also activates the ABA-responsive signalling pathway and other response to biotic stresses. Changes in endogenous ABA levels affect SA-, JA- and ET-related defence responses (Kariola et al. 2006; Asselbergh et al. 2008; Zavala et al. 2009). Incompatible interaction of *Arabidopsis* and *Pseudomonas syringae* is prevented by exogenous ABA treatment (Mohr and Cahill 2007). Several such examples where drought can dramatically affect plant defence responses against pathogens has been reported by Eastburn et al. (2010). Increasing temperatures from 22 to 28 °C reduced the effectiveness of both basal and R gene-mediated resistance in *A. thaliana* when challenged with virulent and avirulent strains of *P. syringae* pv. tomato, respectively (Wang et al. 2009). Increased level of symptoms were found to be the result of changes in the defence responses associated with the host–pathogen interaction, rather than just an increase in pathogen growth at the higher temperature. These studies indicate the need to understand resistance mechanism during interplay between biotic and abiotic stresses in chickpea.

4.5 Conclusions

Changes in climate, with its multiple effects on ecosystems, are likely to change the interactions between an infectious propagule, a susceptible host and favourable environmental conditions leading to the development of new epidemics. The lack of long-term data is hampering the ability to document the certainty changes in disease profiles. For instance in chickpea, surveys and recent investigations clearly indicate the emergence of new diseases like dry root rot and collar rot which has got a direct relationship with temperature and soil moisture. Further, the dynamics affecting host–pathogen interactions is leading to the selection of new pathotypes or pathogens. The changes in temperature and moisture have also shown to affect the disease reaction of cultivars by changing their resistance/susceptible reactions. Therefore, there is a need to address host x pathogen interactions in the light of multiple stress factors. Breeding needs to keep pace with these emerging diseases as it is an essential part of crop improvement. Increases in yield per unit of area will continue to depend largely on more efficient control of stresses (biotic) along with increase in yield potential. Integrated crop management is, therefore, the platform for sustainable agriculture, and extensive research is required in this domain to develop adaptation and mitigation strategies for sustained food security.

Acknowledgement The authors are thankful to the Department of Science and Technology, Climate Change Division for partial funding support provided to conduct various research activities reported in this chapter.

References

Ash GJ, Rees RG (1994) Effect of post-inoculation temperature and light intensity on expression of resistance to stripe rust in some Australian wheat cultivars. Aust J Agric Res 45:1379–1386

Asselbergh B, De Vleesschauwer D, Hofte M (2008) Global switches and fine-tuning-ABA modulates plant pathogen defense. Mol Plant–Microbe Interact 21:709–719

Atkinson NJ, Urwin PE (2012) The interaction of plant biotic and abiotic stresses: from genes to the field. J Exp Bot 63:3523–3543

Brake VM, Pegg KG, Irwin JAG, Chaseling J (1995) The influence of temperature, inoculum level and race of *Fusarium oxysporum* f. sp. *cubense* on the disease reaction of Banana cv. Cavendish. Aust J Agric Res 46:673–685

Dangl JL, Jones JDG (2001) Plant pathogens and integrated defence responses to infection. Nature 411:826–833

Eastburn DM, DeGennaro MM, Delucia EH, Dermody O, McElrone AJ (2010) Elevated atmospheric carbon dioxide and ozone alter soybean diseases at SoyFACE. Glob Chang Biol 16:320–330

FAOSTAT data (2010) http://faostat.fao.org/site/339/default. 2010;aspx. Accessed 12 Nov 2012

French CJ, Elder M (1996) Movement of tomato mosaic virus strains in *Gomphrena globosa* as related to temperature sensitivity and plant resistance. Can J Bot 74:46–50

Ge YF, Johnson JW, Roberts JJ, Rajaram S (1998) Temperature and resistance gene interactions in the expression of resistance to *Blumeria graminis* f. sp. *tritici*. Euphytica 99:103–109

Ghosh R, Sharma M, Telangre R, Pande S (2013) Occurrence and distribution of chickpea diseases in central and southern parts of India. Am J Plant Sci 4:940–944

Gowda CLL, Gaur PM (2004) Global scenario of chickpea research-present status and future thrusts. In: Masood A, Singh BB, Kumar S, Dar V (eds) Pulses in new perspective. Proceedings of the national symposium on crop diversification and natural resource management, Kanpur, 2004, pp 1–22

Graham PH, Vance CP (2003) Legumes: importance and constraints to greater use. Plant Physiol 131:872–877

Harling R, Taylor GS, Matthews P, Arthur AE (1998) The effect of temperature on symptom expression and colonization in resistant and susceptible carnation cultivars infected with *Fusarium oxysporum* f. sp. *dianthi*. J Phytopathol 121:103–117

Hussain A, Iqbal SM, Ayub N, Zahid MA (2006) Factors affecting development of collar rot disease in chickpea. Pak J Bot 38(1):211–216

Hwang SF, Gossen BD, Chang KF, Turnbull GD, Howard RJ, Blade SF (2003) Etiology, impact and control of rhizoctonia seedling blight and root rot of chickpea on the Canadian prairies. Can J Plant Sci 83:959–967

Jimenez-Díaz RM, Alcala-Jimenez AR, Hervas A, Trapero-Casas JL (1993) Pathogenic variability and hosts resistance in the *Fusarium oxysporum* f. sp. *ciceris/Cicer arietinum* pathosystem. In: Proceedings of the 3rd European seminar Fusarium Mycotoxins, taxonomy, pathogenicity and host resistance. Hodowsla Roslin Aklimatyazacja i Nasiennictwo. Plant Breeding and Acclimatization Institute, Radzikov, Poland,,1993., 87e94

Judelson HS, Michelmore RW (1992) Temperature and genotype interactions in the expression of host resistance in lettuce to downy mildew. Physiol Mol Plant Pathol 40:233–245

Jukanti AK, Gaur PM, Gowda CLL, Chibbar RN (2012) Chickpea: nutritional properties and its benefits. Br J Nutr 108:S11–S26

Kariola T, Brader G, Helenius E, Li J, Heino P, Palva ET (2006) Early responsive to dehydration 15, a negative regulator of abscisic acid responses in Arabidopsis. Plant Physiol 142:1559–1573

Kim YK, Bockus WW (2003) Temperature- sensitive reaction of winter wheat cultivar AGSECO 7853 to *Stagonospora nodorum*. Plant Dis 87:1125–1128

Landa BB, Navas-Cortes JA, Jimenez-Díaz RM (2004) Influence of temperature on plant-rhizobacteria interactions related to biocontrol potential for suppression of Fusarium wilt of chickpea. Plant Pathol 53:341–352

Landa BB, Navas-Cortes JA, Jimenez-Gasco MM, Katan J, Retig B, Jiménez-Díaz RM (2006) Temperature response of chickpea cultivars to races of *Fusarium oxysporum* f. sp. *ciceris*, the causal agent of Fusarium wilt. Plant Dis 90:365–374

Mohr PG, Cahill DM (2007) Suppression by ABA of salicylic acid and lignin accumulation and the expression of multiple genes, in Arabidopsis infected with *Pseudomonas syringae* pv. tomato. Funct Integr Genomics 7:181–191

Nawaz KS (2007) *Macrophomina phaseolina* as causal agent for charcoal rot of sunflower. Mycopath 5:111–118

Olaya G, Abawi GS (1996) Effect of water potential on mycelial growth and on production and germination of sclerotia of *Macrophomina phaseolina*. Plant Dis 80:1347–1350

Omwega CO, Roberts PA (1992) Inheritance of resistance to Meloidogyne spp. in common bean and the genetic basis of its sensitivity to temperature. Theor Appl Genet 83:720–726

Pandey P, Sinha R, Mysore KS, Senthil-Kumar M (2014) Impact of concurrent drought stress and pathogen infection on plants. In: Mahalingam R (ed) Combined stresses in plants: physiological, molecular, and biochemical aspects. Springer, Cham, pp 203–222

Pearson CAS, Schwenk FW, Crowe FJ (1984) Colonization of soybean roots by *Macrophomina phaseolina*. Plant Dis 68:1086–1088

Raabe RT (1985) Association of soil particles with seeds and three pathogens of chickpea in California. Indian Phytopathol 69:238–239

Ryan J (1997) A global perspective on pigeon pea and chickpea sustainable production systems: present status and future potential. In: Asthana A, Ali M (eds) Recent advances in pulses research. Indian Society for Pulses Research and Development, Kanpur, pp 1–31

Savary S, Nelson A, Sparks AH, Willocquet L, Duveiller E, Mahuku G, Forbes G, Garrett KA, Hodson D, Padgham J, Pande S, Sharma M, Yuen J, Djurle A (2011) International agricultural research tackling the effects of global and climate changes on plant diseases in the developing world. Plant Dis 95:1204–1216

Sharma M (2012) Climate change, emerging plant diseases and their management: recent developments. In: Global meet of Biologists, Hyderabad, December 26–28, pp 43–45

Sharma M, Pande S (2013) Unravelling effects of temperature and soil moisture stress response on development of dry root rot [*Rhizoctonia bataticola* (Taub.)] butler in chickpea. Am J Plant Sci 4:584–589

Sharma M, Ghosh R, Pande S (2015) Dry root rot (*Rhizoctonia bataticola* (Taub.) Butler): an emerging disease of chickpea – where do we stand? Arch Phytopathol Plant Protect 48(13–16):797–812

Singh G, Sharma YR (2002) Fungal diseases of pulses. In: Gupta VK, Paul YS (eds) Diseases of field crops. Indus publishing, New Delhi, pp 155–192

Sydenham GM, McSorley R, Dunn RA (1997) Effects of temperature on resistance in *Phaseolus vulgaris* genotypes and on development of *Meloidogyne* species. J Nematol 29:90–103

Varshney RK, Mohan SM, Gaur PM, Gangarao NVPR, Pandey MK, Bohra A, Sawargaonkar SL, Chitikineni A, Kimurto PK, Janila P, Saxena KB, Fikre A, Sharma M, Rathore A, Pratap A, Tripathi S, Datta S, Chaturvedi SK, Mallikarjuna N, Anuradha G, Babbar A, Choudhary AK, Mhase MB, Bharadwaj C, Mannur DM, Harer PN, Guo B, Liang X, Nadarajan N, Gowda CL (2013) Achievements and prospects of genomics-assisted breeding in three legume crops of the semi-arid tropics. Biotechnol Adv 3:1120–1134

Wan J, Dunning M, Bent AF (2002) Probing plant-pathogen interactions and downstream defense signalling using microarrays. Funct Integr Genomics 2:259–273

Wang Y, Bao Z, Zhu Y, Hua J (2009) Analysis of temperature modulation of plant defense against biotrophic microbes. Mol Plant–Microbe Interact 22:498–506

Zavala JA, Casteel CL, Nabity PD, Berenbaum MR, DeLucia EH (2009) Role of cysteine proteinase inhibitors in preference of Japanese beetles (*Popillia japonica*) for soybean (*Glycine max*) leaves of different ages and grown under elevated CO_2. Oecologia 161:35–41

Genomics-Assisted Breeding for Improving Stress Tolerance of Graminaceous Crops to Biotic and Abiotic Stresses: Progress and Prospects

5

Roshan Kumar Singh, Pranav Pankaj Sahu, Mehanathan Muthamilarasan, Annvi Dhaka, and Manoj Prasad

Abstract

Advances in genomics research have led to the development of high-quality reference genome data, genome-wide molecular markers, quantitative trait loci (QTL), and high-throughput genotyping platforms for cereal crops. The availability of these genomic resources has facilitated the development of breeding technologies such as genomics-assisted breeding (GAB). GAB is an advanced form of marker-assisted breeding where genome-wide genetic selection and high-density genotyping are performed to generate elite varieties with better agronomic traits. Marker-assisted selection (MAS) is a genotypic variation based indirect selection method that reduces the time and cost of breeding. The different approaches of MAS include marker-assisted backcrossing (MABC) or introgression of agronomically important alleles or QTLs with relatively large effect, marker-assisted recurrent selection (MARS) for introduction of complex traits and genomic selection (GS) based on overall molecular markers distributed throughout the genome. In view of these, the present chapter discusses the application of genetic and genomic resources in identification and mapping of stress-tolerant genes/QTLs and their application in molecular breeding. In addition, the chapter also summarizes the current status of marker-assisted selection approach for improving tolerance to drought and virus infection in major graminaceous crops. The challenges and future prospects of GAB in enhancing crop productivity under stress conditions have also been summarized.

Keywords

Marker assisted selection • Millets • Drought • QTLs • Yield traits • Crop production • Cereal crops

---

R.K. Singh • P.P. Sahu • M. Muthamilarasan • A. Dhaka • M. Prasad (✉)
National Institute of Plant Genome Research, Aruna Asaf Ali Marg, New Delhi 110 067, India
e-mail: manoj_prasad@nipgr.ac.in

M. Senthil-Kumar (eds.), *Plant Tolerance to Individual and Concurrent Stresses*,
DOI 10.1007/978-81-322-3706-8_5

5.1 Introduction

Gramineae or Poaceae is one of the largest plant families comprising of monocotyledonous grasses. All the major cereals such as rice, wheat, and maize and minor cereals including millets, sorghum, and *Brachypodium* sp. (hereafter referred as brachypodium) are classified under this family, and of note, these crops feed more than 90% of world's population. The subfamilies, Pooideae and Panicoideae, diverged from a common Poaceae ancestor around 70 million years ago (mya), and it includes brachypodium and wheat in the former subfamily, whereas millet, sorghum, and maize are included in the latter. The subfamily Ehrhartoideae diverged from Pooideae ~34 mya, and this includes the major cereal crop, rice. Being served as important food and feed crops, cereals have been extensively domesticated around the world, and their domestication has a close relation to human civilization. Till now, the global population depends on these cereals for food and feed, and, therefore, the importance of cultivating cereals is increasing day by day. This has given rise to many advanced technologies, strategies, and practices, which are being followed by the farmers to improve the productivity. However, the emission of greenhouse gases due to increased industrial activities has a drastic effect on global climate change, which in turn poses a serious threat to agriculture. Poaceae members with C_3 photosynthesis mechanism (including rice, wheat, and brachypodium) are highly susceptible to the consequences of climate change. Conversely, the C_4 crops (millet, sorghum, and maize) perform comparatively better as they have relatively higher water use efficiency and nitrogen use efficiency. Therefore, C_4 crops sustain best in arid and semiarid regions irrespective of soil nutrition, improper water supply, and supply of fertilizers, while C_3 crops are being cultivated in well-irrigated landscapes as they are highly susceptible to any change in climate, temperature, irrigation cycle, and cultivation period. However, all the major cereals fall under C_3 category, and their productivity needs to be ensured/enhanced to feed the growing population, which is expected to reach nine billion by 2050 (Muthamilarasan et al. 2013).

The key challenges associated with improving the productivity of major cereals are the consequences of climate change including increase in temperature, decrease in groundwater level, severe drought, higher salinity, and poor soil nutrition. These adverse consequences are termed "stresses" as they challenge the survival, growth, and productivity of crops. Among these stresses, water scarcity (drought) is the immediate outcome of global warming, which will be associated with other stressors in a combinatorial mode, thus posing serious threat to crop survival and productivity. In view of this, improving the drought tolerance of major cereals has been a prime concern of plant biologists. Both transgene-based and molecular breeding approaches have been implied to enhance the tolerance of cereals to drought and other stresses; however, breeding for stress tolerance has gained importance owing to its much debated "safety." Advancements in next-generation sequencing (NGS) technologies have led to the development of large-scale, genome-wide novel markers, alleles, SNPs, and trait-associated QTLs in major as well as orphan (less studied) crops that were neglected previously. The knowledge of molecular makers, functional alleles, and QTLs that confers desirable traits (e.g., drought tolerance) help researchers to develop new varieties that combine those traits without compromising yield potential. Thus, the advanced breeding strategies that utilize genomics information are referred as genomic-assisted breeding (GAB). GAB is a modern breeding technology that integrates structural, functional, and comparative genomics to identify functional molecular markers, QTLs, candidate genes, and predictive markers for breeding (Varshney et al. 2015).

Unlike GAB for drought tolerance, breeding for virus resistance has invited less research attention among the plant biologists, and the advent of NGS and high-throughput analysis platforms have not made much progress in the area of advanced molecular breeding for virus resistance. In view of this, the present chapter provides a comparative note on the progress and prospects of GAB for drought tolerance and virus resistance in major cereals.

5.2 Genomic Resources and Their Applications in Genomic-Assisted Breeding for Drought Tolerance

5.2.1 Molecular Markers

Molecular markers allow the identification of genetic variation among cultivars and provide a powerful tool to breeders for identifying the appropriate parents for crosses and to select most desirable individual among offspring of the cross. Among different types of molecular markers, microsatellite or simple sequence repeat (SSR) and single nucleotide polymorphism (SNP) have been routinely used in plant breeding programs (Gupta and Varshney 2000; Appleby et al. 2009). SSR marker represents polymorphism in terms of number of repeat units, and they have proven convenient in molecular breeding since they show codominant inheritance, multiallelic nature, genome specificity, and abundance throughout the genome (Gupta and Varshney 2000; Varshney et al. 2005; Ganal and Roder 2007). They have widely been used in gene tagging, and identification and analysis of QTLs associated with the traits, localization of genes and QTLs to chromosomes, and genomic-assisted selections. In view of their importance, large-scale mining of microsatellite markers has been reported in cereals. Zhang et al. (2007) identified 102,706 and 102,423 microsatellite markers in indica and japonica rice varieties, respectively. These microsatellites predominantly cover the intergenic regions (56.6% for indica and 57.4% for japonica) as compared to intronic regions (20.3% for indica and 19.2% for japonica). In another study, 19,555 rice genic noncoding microsatellite (GNMS) repeats have been identified and validated with an average of 357.5 GNMS repeats per Mb of genome (Parida et al. 2009). Han et al. (2015) have identified 364,347 genome-wide microsatellite markers based on 10,603,760 sequences of Chinese spring wheat genome with a density of 36.68 microsatellite markers per Mb. They have detected 488 types of microsatellite motifs with the proportion of 42.52% of dinucleotide, 24.94% of trinucleotide, 4.62% of tetranucleotide, 3.25% of pentanucleotide, and 24.65% of hexanucleotide repeats of the genome.

Whole-genome analysis has led to the identification of 264,658 SSRs from 17 maize varieties (Xu et al. 2013a). In another study, Qu and Liu (2013) have identified 179,681 SSRs from the B73 cultivar of maize. In a single barley genetic map, Zhou et al. (2015) have developed 1140 InDel markers integrated with 383 SSRs, 3909 gene-based SNPs, and 1544 DarT (diversity arrays technology) markers by aligning genomic DNA sequences of Morex and Barke barley cultivars. A genome-wide analysis has identified 28,324 SSR motifs covering 405.3 Mb of *Setaria italica* genome with an average of about 69 SSR motifs per Mb of genome sequence (Pandey et al. 2013). A set of 22,879 microsatellite markers were identified and mapped genome of brachypodium with a density of 101 SSR markers per Mb (Sonah et al. 2011). A set of 365 EST-SSRs with trinucleotide motifs were identified of which 287 were highly polymorphic among 18 genotypes of sugarcane (Marconi et al. 2011).

Another important class of molecular markers is single nucleotide polymorphisms (SNPs) that are more abundant and widely distributed in the genome and amenable to automation for high-throughput genotyping (Varshney et al. 2006; Mammadov et al. 2012). SNP identification in crops is a difficult task because of the genome complexity and often lacks reference genome data. SNP discovery depends on comparative genome sequencing of varieties or analysis of expressed sequence tags (ESTs) from different lines (Habash et al. 2009). Recent advances in next-generation sequencing (NGS) technologies such as 454/flx, Illumina Solexa, and SOLiD have been used to identify large-scale SNPs throughout the genomic regions in a number of cereals. Alexandrov et al. (2015) have identified about 20 million SNPs in rice by aligning 3000 rice genome sequences with the reference genome of Nipponbare. In rice, 162,380 genome-wide insertion-deletion polymorphism markers with polymorphism information content (PIC) $\geq$0.5 have been identified using NGS technology (Liu et al. 2015a). A total of 46,977 gene-associated SNPs has been genetically mapped on wheat genome using a high-density 90,000 SNP array (Wang et al. 2014a). In maize, 6,305,011 SNPs were identified from 15 different drought-tolerant and drought-susceptible inbred lines by resequencing using Illumina Hiseq 2000 platform (Xu et al. 2014). Based on this data, non-synonymous SNPs (nsSNPs) and 271 nsSNP-associated drought stress-regulated candidate genes were identified. A total of 283,000 SNPs were discovered in sorghum by high-throughput sequencing of eight diverse cultivars (Nelson et al. 2011). In barley, approximately 22,000 SNPs were detected from ESTs and sequenced amplicons, and among these, 4596 were genotyped for performance using Illumina GoldenGate assay in three pilot phases (Close et al. 2009).

5.2.2 Transcriptome Analysis

Before whole-genome sequencing approach was introduced, a large number of transcriptome sequencing projects were accomplished in several cereal crops. High-throughput comparative transcriptome analysis has been used to identify genes expressed specifically in response to drought condition. These studies have proven valuable in understanding gene function and the molecular basis of different cellular processes even in the absence of complete genome sequence data (Varshney et al. 2015). Comparative transcriptome studies reveal that a large number of diverse group of genes and pathways have involved in sensing and responding to drought stress in plants (Habash et al. 2009). A number of candidate genes responsible for tolerance to water stress have been identified and cloned from several crop plants (Sehgal and Yadav 2010).

In recent years, advanced NGS technologies have been using for analyzing the gene expression patterns via de novo transcriptome assemblies. These transcriptome assemblies have led to the discovery of large number of functional molecular markers linked to drought specific genes. Comprehensive drought-associated differential gene expression profiling has been performed in rice using global transcriptome sequencing of tolerant introgression line and its parent lines (Huang et al. 2014). Global wheat transcriptome analysis under heat- and drought-stressed samples through Illumina Hiseq2000 has identified 29,395 genes that were differentially expressed in at least single stress condition (Liu et al. 2015b). In maize, a genome-wide transcriptional analysis from normal and abiotic stress-treated tissues resulted in generation of 27,455 full-length cDNAs which were sequenced, mapped, and analyzed (Soderlund et al. 2009). Transcriptome sequencing of drought-tolerant and drought-susceptible cultivars of barley using a 454 GS FLX Sequencer revealed 800 unique transcripts and 1017 SNPs in these two ecotypes (Bedada et al. 2014). A whole-genome transcriptome database, "MOROKOSHI," was constructed harboring 37,607 full-length cDNA clones and their sequencing by Sanger method to obtain 38,981 ESTs in sorghum. Reference-based and de novo transcriptome analyses of *Setaria viridis* have led to the identification of 42,754 and 60,751 transcripts, respectively, using Illumina Hiseq 2000 sequencing platform. From these transcripts, 7056 and 9576 EST-SSRs have been identified (Xu et al. 2013b). Genome sequence of *S. italica* was used as reference genome for generating these transcripts and SSRs. The mRNA library of three genotypes of *Brachypodium sylvaticum* was sequenced on the Illumina HiSeq 2000 platform to produce more than 350 million reads, which were aligned with *B. distachyon* Bd21 reference genome (Fox et al. 2013). Upon aligning sequence reads of *B. sylvaticum* against *B. distachyon* Bd21, 394,654 SNPs and >20,000 SSRs were identified. The de novo transcriptome assembly of sugarcane was produced using Illumina RNA-Seq platform with more than 400 million reads, which revealed 72,269 unigenes, 708,125 SNPs, and 5106 SSRs (Cardoso-Silva et al. 2014).

5.2.3 Functional and Comparative Genomics

Functional genomics involves the study of gene function and interaction among various genes and their regulation to deliver a biological function in an organism. The main focus of functional and comparative genomics is to identify the functional allelic differences which are responsible for improved phenotype (Varshney et al. 2007). Gene function and their regulations are defined by the study of transcription, translation, and interaction of the genes with other genes (Kumpatala et al. 2012). Several tools and techniques such as quantitative PCR (qPCR), micro- and macroarrays, serial analysis of gene expression (SAGE), massively parallel signature sequencing (MMPS), and RNA-Seq are routinely used for expression analysis of large number of genes. qPCR utilizes fluorescent probe in the reaction mixture to determine accurate quantification of transcript in the given sample. Two of more genes can be quantified in the same PCR reaction using different fluorescent probes or dyes. The fluorescence emission of the dye increases relatively with the accumulation of target gene product in the PCR reaction.

Comparative analysis of gene expression in a large set of genes has been made possible by microarray and macroarray technologies. Affymetrix rice gene chip array containing 4856 japonica and 1360 indica sequences has been used to detect 5284 genes that were differentially expressed under drought stress condition (Wang et al. 2011). Differential gene expression through microarray-based technique in wheat with divergence levels of transpiration efficiency under drought conditions has been performed, and 93 genes were identified, in which most of them were directly associated with drought stress (Xue et al. 2006). Microarray-based gene expression analysis was studied in two drought-tolerant and drought-resistant cultivars of maize, and it was found that a large number of genes were differentially expressed in response to drought (Hayano-Kanashiro et al. 2009). These differential gene expression patterns have suggested that drought-tolerant cultivars have a mechanism to induce a variety of regulator genes under water-stress conditions that modulates a wide range of metabolic and cellular responses under drought. Differential gene expression analysis followed by quantitative real-time PCR has shown that genes associated with drought stress-responsive pathways such as abscisic acid, jasmonic acid, and phenylalanine ammonia-lyase have been upregulated during onset of drought in tolerant variety (Luo et al. 2010). Combinations of different abiotic stresses produce a completely different gene expression pattern than the individual stress alone (Humbert et al. 2013). Transcriptome analysis of individually heat and drought-stressed sorghum along with combined stressed plants revealed gene expression pattern which varied significantly in each case. However, common expression of some genes in combined stresses has shown evidence for cross talk between the stresses (Johnson et al. 2014). Affymetrix Barley1 microarray was used to detect differentially expressed genes in two drought-tolerant barley ecotypes, Martin and *Hordeum spontaneum* 41–1 (HS41–1), and one drought-sensitive genotype Moroc9–75 under drought conditions (Guo et al. 2009). Seventeen genes were identified, expressed only in tolerant ecotype and not in susceptible one, of which 12 genes were highly expressed at all time points. An analysis of global gene expression under different abiotic stresses in *B. distachyon* has been performed using microarray technology, which revealed significant downregulation of 40, 1621, 1137, and 5790 genes in response to cold, heat, salt, and drought stresses, respectively. In contrast, 447, 458, 1565, and 2290 genes were significantly upregulated in response to cold, heat, salt, and drought stress, respectively (Priest et al. 2014).

SAGE and MMPS are powerful tools for analysis of gene expression based on generation of short sequence tags of 12–18 bases from cDNA libraries. The advantage of SAGE and MMPS over microarray is its simultaneous de novo detection of all mRNA in the given sample, while microarrays are limited to genes fixed on the chip. Previously, the obtained tags from both the methods were used to ligate to each other for generating long stretches of tags and then sequenced by Sanger sequencing. Recently, with the advancement with NGS technology, these tags can be isolated and sequenced individually to provide absolute quantitative gene expression information. Typical SAGE dataset comprises of 40,000–60,000 tags, while MMPS comprises of millions of signature sequences. The major drawback of SAGE and MMPS is that these techniques require large number of tags and can be very expensive to analyze and screen.

5.3 Mining of QTLs Associated with Drought Tolerance Traits

It has been reported that drought tolerance is a complex quantitative trait regulated by a number of minor genes or small main-effect QTLs (Barnabas et al. 2008; Ravi et al. 2011). Wild varieties of many C_4 panicoids have been proven to be the genetic sources of drought tolerance alleles or QTLs, which have transferred to the high-yielding local varieties through marker-assisted selection (MAS) in order to enhance grain yield under drought conditions (Tuberosa and Salvi 2006). This method of transferring of trait-associated QTLs from wild relatives into a cultivated crop variety is commonly called advanced backcross QTL analysis (Tanksley and Nelson 1996). Linkage analysis-based QTL mapping has been the most common approach during the last decade for QTL mining and implemented in a number of crops to find QTLs associated with drought resistance traits (Mir et al. 2010); however, recently linkage disequilibrium (LD)-based association mapping has been developed as an alternative to linkage analysis for QTL mapping for dissecting complex abiotic and biotic traits (Myles et al. 2009; Mir et al. 2010; Maccaferri et al. 2014).

The interaction of G×E (genotype versus environment) confers serious problems in transferring QTLs from wild to cultivated crops. One of the major limitations of QTL study is its environmental dependent expression, i.e., often QTL established in one environment may be absent in another (Habash et al. 2009). To understand the interaction of G×E, multi-locational field trail of the same population is necessary, and the impact of environment on particular genotype can be analyzed. The multi-locational phenotyping trail helps in identification of stable QTLs that remain unchanged across the environment (Vargas et al. 2006). Messmer et al. (2009) have identified a number of stable QTLs related to drought tolerance in recombinant inbred line (RIL) population of maize through QTL-by-environment interaction (QTL × E) study across several locations under different levels of irrigation. The identified QTLs may prove valuable for the breeders to design drought-tolerant high-yielding corn varieties. QTLs identified for drought tolerance and increase in grain yield in cereal crops are summarized in Table 5.1.

Table 5.1 Summary of QTLs identified using GAB for drought tolerance and grain yield in major graminaceous crops

Crop	Trait	Chromosome/ linkage group	Phenotypic variation explained (PVE %)	References
Rice	Yield under drought stress	1, 4, 6	9.9–20.9	Prince et al. (2015)
	Grain yield under drought	2, 3	13.0–31.0	Venuprasad et al. (2009)
	Drought tolerance	2,4,5,6,7,8	10.0–22.0	Kato et al. (2008)
	Grain yield under drought conditions	All chromosomes except 12	7.5–55.7	Lanceras et al. (2004)
	Drought stress and stress indicator	All chromosomes except 5	5.0–59.0	Babu et al. (2003)
	Drought avoidance	All chromosomes except 9	4.4–25.6	Price et al. (2002)
Wheat	Potential quantum efficiency of photosystem II, chlorophyll content, flag leaf temperature, and grain yield	1A, 1D, 2B, 3A, 3B, 4B, 4D, 5B, 6A	16.3–37.0	Kumar et al. (2012)
	Drought resistance	All 14 chromosomes	0.8–42.4	Peleg et al. (2009)
	Heat and drought adaptation	All except 2A, 2D, 3D, 5D, 6D, and 7D	3.1–13.2	Pinto et al. (2010)
	Drought tolerance	1B, 1D, 2B, 3A, 3B, 4A, 4B, 4D, 5A, 5B, 6A, 6B, 6D, 7A, 7B	-	Mathews et al. (2008)
Maize	Drought tolerance	1, 3, 4, 5, 7, 10	1.6–19.5	Almedia et al. (2013)
	Grain yield and anthesis silking interval under drought and well-watered conditions	All chromosomes	1.2–13.1	Semagn et al. (2013)
	Drought tolerance	1, 3, 5, 6, 7, 9	0.2–75.0	Rahman et al. (2011)
	Grain yield and associate trait under drought condition	1, 4, 5, 6, 8, 9	1.68–13.3	Guo et al. (2008a)
	Drought tolerance and grain yield	1, 2, 3, 5, 7, 8, 9	4.1–31.3	Xiao et al. (2005)
Barley	Yield traits under drought conditions	All chromosomes	6.5–36.9	Von Korff et al. (2008)
	Chlorophyll and chlorophyll fluorescence parameter	1H, 2H, 4H, 6H, 7H	6.2–13.6	Guo et al. (2008b)
	Drought-related traits	1H, 2H, 3H, 4H, 5H, 6H, 7H	4.0–16.0	Diab et al. (2004)

(continued)

Table 5.1 (continued)

Crop	Trait	Chromosome/ linkage group	Phenotypic variation explained (PVE %)	References
Sorghum	Grain yield under drought conditions	All chromosomes	2.5–30.3	Nagaraja et al. (2013)
	Nodal root angle and drought-associated traits	SBI-01, SBI-02, SBI-05, SBI-08, SBI-10	23–58.2	Mace et al. (2012)
	Drought tolerance	2H, 3H, 4H, 5H, 6H, 8H, 9H	-	Sabadin et al. (2012)
Pearl millet	Drought tolerance	2	32	Yadav et al. (2011)

5.4 Genomic-Assisted Breeding (GAB) for Drought Tolerance

The major objective of crop improvement program is to develop various abiotic, biotic, or combined stress tolerance and high-yielding varieties from the existing elite and wild varieties. The genomics approach aids better understanding of molecular mechanism of plant responses under various stress conditions and allows selection of desired progeny of cross at very early stage of development which was missing in conventional breeding. The basic principle of GAB is to identify molecular marker associated with the trait(s) of interest that helps in early selection of desired progeny in breeding cycle by the use of high-throughput genotyping platforms (Varshney et al. 2013). Several marker-assisted selection (MAS) schemes have been used in GAB which include marker-assisted backcrossing (MABC), marker-assisted recurrent selection (MARS), and genomic selection (GS).

5.4.1 Marker-Assisted Backcrossing (MABC)

Marker-assisted backcrossing is the simplest and widely used form of MAS, which involves introgression of a genomic locus (gene or QTL) associated with trait of interest from a donor parent into an elite cultivar or breeding line through several generations of backcrossing. The desired product of MABC is a breeding line having whole genome of recurrent parent with the desired trait (Fig. 5.1). With the help of molecular markers, breeders may identify the individuals that have the desired trait with highest percentage of recurrent parent genome. Selection can be either of foreground or of background (Hospital and Charcosset 1997). In foreground selection, plants having allele-specific markers to the donor parent at the target locus are selected. The selected individuals which are having desired locus at heterozygous state after several round of backcrossing are self-pollinated and progenies identified having target locus as homozygous condition. In background selection, plants having marker allele specific to the recurrent parents throughout

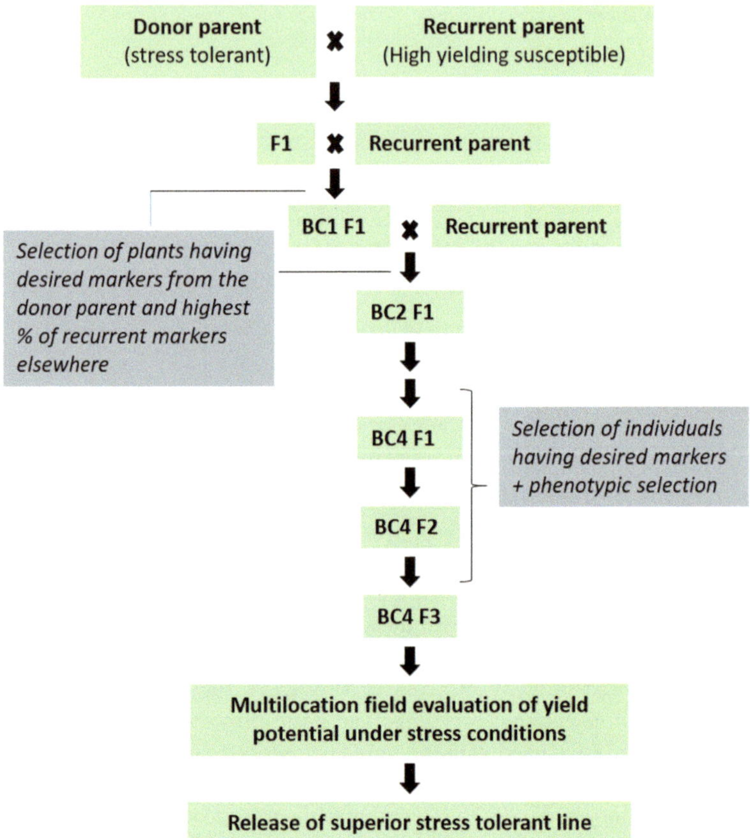

Fig. 5.1 Marker-assisted backcrossing for development of stress-tolerant crop plants

the genomic region except the target locus are selected. The efficiency of MABC depends on population size for each generation of backcrossing, number of markers used for the trait, marker and target gene association, and undesirable linkage drag.

Drought-tolerant near isogenic lines (NILs) of rice have been developed by introgression of three root QTLs from CT9993, a drought-tolerant japonica upland cultivar into IR20, a lowland drought-sensitive indica cultivar following MABC (Suji et al. 2012). NILs representing the root QTLs were high grain yielding under drought conditions. Maize is the most studied C_4 panicoid grain crop in which MABC was conducted for generating drought-tolerant breeding lines. For example, MABC selection experiment was performed at CIMMYT, Mexico, for improvement of grain yield under drought conditions in tropical corn. The desired breeding line was developed by crossing of drought-susceptible line CML247 (recurrent parent) with drought-resistant line Ac7643 (donor parent) followed by marker-assisted selections of successive four generations from BC_1F_1 to BC_2F_2. The phenotyping of selected genotypes was performed under low-water conditions, and it was confirmed that the resultant lines produce more grain yield than recurrent parent under drought (Ribaut and Ragot 2007). In the case of sorghum, drought-tolerant variety was developed by marker-assisted introgression of stay-green QTL region from a drought-tolerant donor variety into adapted local farmer variety (Ngugi et al. 2013). The donor parent, E36–1, was backcrossed into Kenya local farmer-preferred variety, Ochuti, and foreground and background selection were performed to select the individuals having highest proportion of Ochuti genome with stay-green QTL region from E36–1 line. They have estimated that the genotype where stay-green region has transferred grows better under water-limited condition. The application of MABC has also been applied for development of drought-tolerant pearl millet. A major QTL responsible for high grain yield under water-deficient conditions has been identified and transferred from a donor inbred line PRLT 2/89–33 into a drought-susceptible cultivar H 77/8–332 of pearl millet through MABC (Serraj et al. 2005).

5.4.2 Marker-Assisted Recurrent Selection (MARS)

The major limitation in MABC is its inefficiency to transfer QTLs for complex traits governed by multiple genes or large QTLs. This restriction could be overcome by marker-assisted recurrent selection (MARS), which includes identification and selection of multiple genomic regions for a complex trait using molecular makers within a single or across related populations (Bernardo 2008; Ribaut et al. 2010). This scheme is effective in improving quantitative traits in cross-pollinated species like maize, sunflower, sorghum, rice, wheat, etc. In contrast to MABC, favorable alleles can be contributed by both the parents in this method and the outcome results in a genotype containing genetic regions of both the parents. MARS involves identification and selection of desired traits from F2 population based on both phenotypic data and marker effects followed by two or three recombinational cycles of marker-based selection only (Fig. 5.2).

In maize MARS programs, a large-scale use of markers in biparental populations has been demonstrated. Initially, the markers were utilized for QTL detection and then applied for MARS on yield (viz., rapid cycles of recombination and selection based on associated markers for yield). This enhanced the efficiency of long-term selection by increasing the frequency of favorable alleles (Jhonson 2004). Eathington et al. (2007) and Crosbie et al. (2006) also demonstrated that the genetic gain achieved through MARS in maize was double that of phenotypic selection (PS) in a few reference populations.

5.4.3 Genomic Selection (GS)

Genomic selection or genome-wide selection (GWS) is a form of MAS, which uses whole-genome molecular markers (high-density markers throughout the genome) for improving quantitative traits in plant breeding program (Meuwissen et al. 2001).

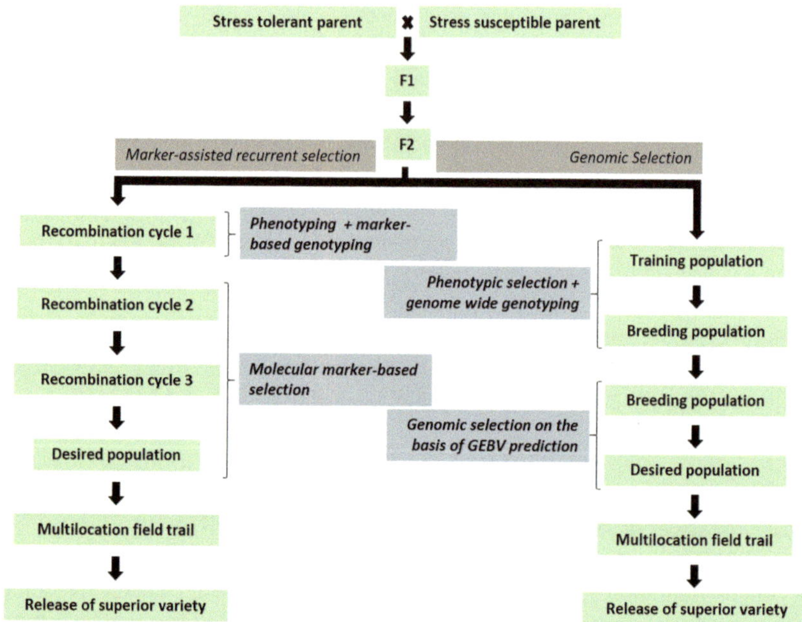

Fig. 5.2 Marker-assisted recurrent selection and genomic selection for development of stress-tolerant crop plants

Phenotyping is not essential for breeding population, selection being based on genomic predictions that combine genotypic and pedigree data over several generations for population in order to increase the accuracy of prediction (Nakaya and Isobe 2012). The selection of desired genotype is based on genomic estimated breeding value (GEBV) which is information index of combined effect of genome-wide DNA markers (Fig. 5.2). In the first phase of GS (training phase), phenotype and genome-wide genotype using large number of dense molecular markers of training population (breeding lines used in breeding program) are investigated for overall performance across the environment to predict the significant relationship between genotypes and phenotypes by means of statistical approach. Subsequently, GEBVs are applied for the selection of superior individuals for next crosses in the breeding phase. Thus, in GS, data obtained from GEBV estimates are used for selection of desired individuals, instead of genotypes of molecular markers used in traditional MAS. In breeding phase, phenotyping is not required for selection; however, thorough genome-wide genotyping over several generations for all the individuals are required. High molecular marker density in which all quantitative trait loci (QTLs) are in linkage disequilibrium with at least one marker is essential for accurate GEBV and GS (Habier et al. 2007).

The application of GS in molecular breeding has reduced the cost and time of breeding by decreasing the number of selection events. Bernardo and Yu (2007) conducted a comparative study of GS and MARS in maize breeding program and concluded that genome-wide selection is superior over MARS for complex traits without having subset of markers with significant effect. In another example, it has been shown that genome-wide selection for drought resistance in high-yielding variety of maize has proven advantageous than the indirect phenotypic selection through secondary traits (Ziyomo and Bernardo 2013). The major advantage of GS over other forms of selection methods is that it decreases time period of breeding by reducing the number and frequency of phenotyping and cost-effective as genotype-based selection is much cheaper than the phenotypic selection (Varshney et al. 2013; Xu et al. 2012; Ziyomo and Bernardo 2013).

5.5 Genomics-Assisted Breeding for Improving Virus Tolerance

The NGS and genotyping technologies have also emerged as key tools to revolutionize plant breeding and crop improvement programs for virus resistance. Progresses in sequencing and genotyping technologies have enabled the researchers to generate and exploit available comprehensive high-quality genomic resources such as genomes, molecular markers, and BAC-end sequences to reduce losses due to virus infection in cereals. This advancement has also supported the genomics-driven breeding in providing solution to the problems of limited genetic improvement and low productivity in graminaceous crops due to virus infection. In this section, we have summarized the present scenario of crop yield loss due to virus infection and prospects for GAB in improving tolerance in economically important graminaceous crops.

Maize is the natural host of several viruses, and QTLs linked with resistance to *Sugarcane mosaic virus* (Xia et al. 1999; Zhang et al. 2003), *Maize mosaic virus* (Zambrano et al. 2014), *Wheat streak mosaic virus* (McMullen and Simcox 1995; Jones et al. 2011; Zambrano et al. 2014), *Maize dwarf mosaic virus* (Zambrano et al. 2014), and *Maize chlorotic dwarf virus* (Jones et al. 2011) have been reported in maize. Although most of the studies were based on conventional breeding approaches, genomics-assisted breeding has also been applied to identify trait loci related to virus resistance in maize. Recently, QTL for BYDV resistance in maize was identified through GWAS (Horn et al. 2014), which revealed the presence of SNPs on chromosome 4 and 10 in a core collection of maize germplasms, which explained a high level of phenotypic variation for traits linked with BYDV resistance. Recently, high-density SNPs were identified in maize, and a GWAS was performed to define the genetic architecture linked with maize rough dwarf disease (MRDD) resistance (Tao et al. 2013). Whole-genome SNP analysis revealed that 14.2% of SNPs differed between contrasting lines of maize (susceptible, NT409; resistant, NT411) varied in MRDD resistance. Through this, a *qMrdd1* locus linked with MRDD resistance was fine mapped in maize genome (Tao et al. 2013). Further, study of *qMrdd1* locus resulted in identification of genes with high similarity to pathogenesis-related (PR) proteins including Jumonji C domain (GRMZM2G417089), and a gene encoding ethylene pathway-specific transcription factor in activating defense (GRMZM2G055204; Chen et al. 2015). MaizeSNP50 DNA analysis tool has assisted in the interrogation of genetic variation across maize lines.

Recently, a genome-wide physical map of barley has been generated through fingerprinting of ~600 BAC clones representing 14-fold haploid genome coverage. This would certainly be helpful in identifying and fine mapping virus resistance-associated QTLs in barley. Interestingly, 436,640 InDels have been identified while aligning the two barley cultivars (Morex and Barke), out of which 383 SSRs, 3909 gene-based SNPs, and 1544 DArT markers were further integrated into single barley genetic map (Zhou et al. 2015). Barley cultivation in Europe and East Asia is susceptible to diseases caused by *Barley yellow mosaic virus* (BaYMV) and *Barley mild mosaic virus* (BaMMV). Report is available on the centromeric specific recessive resistance gene *rym11* associated with BaMMV and BaYMV at chromosome 4HL. The available barley EST sequences along with the NGS data of barley were exploited, and a marker co-segregating with *rym11* was identified for efficient marker-assisted selection (Lüpken et al. 2013). Furthermore, cross between *rym11* genotypes × cultivar carrying the recessive resistance gene *rym1* was done, and diagnostic PCR-based markers were developed to differentiate known resistance-conferring alleles of the *rym11* locus. Moreover, breeding tools have been developed for MAS of rym11 in barley breeding (Yang et al. 2014).

Brachypodium distachyon has not only emerged as a model species for the study of many cereal crops such as barley, wheat, oats, and rye but also has been established as a model plant to study plant-pathogen interaction (Fitzgerald et al. 2015). The most important breakthrough in *Brachypodium* research was the release of a high-quality draft genome sequence for inbred line Bd21 (The International Brachypodium Initiative 2010). This advancement has helped in the generation of resources to efficiently fine map the associated loci of *B. distachyon* controlling various agronomic traits including virus resistance. Several strains of *Barley stripe mosaic virus* (BSMV) have been reported to infect *B. distachyon* (Petty et al. 1994). An attempt to fine map the BSMV resistance locus has been performed through utilizing genetic linkage map of RIL population using SNP markers. It was revealed through high-throughput genomic sequencing that 23 kbp region of chromosome 3 was linked with the *Bsr1* locus (Cui et al. 2012). Furthermore, it has been highlighted that the corresponding 23 kb region has specific ORFs related to plant defense; for example, NB-ARC and LRR domains containing R genes (Cui et al. 2012). Hence, the availability of draft genome of *Brachypodium* has helped the plant breeders to apply the genomics-assisted breeding to define which plant gene(s) are keys to virus resistance.

Genomics-assisted breeding has a significant impact on rice breeding. Exploitation of these genomics resources for the identification of virus resistant traits, superior examination of useful genetic variation, and rapid transfer of loci/genes in rice will be the key factor for breeding programs. In this context, 3000 Rice Genomes Project has been initiated to generate a giga-dataset of genome sequences derived from a large set of 3000 global accessions of rice (Li et al. 2014). These publicly available dataset has assisted the breeders to identify the regions associated with virus resistance. Interestingly, out of these 3000 rice accessions, 5.3% were found to be the potentially source of resistance to at least one of the rice-infecting viruses.

Around 16 species of plant viruses have been reported to cause disease in rice (Zhou et al. 2013). Among them, *Rice yellow mottle virus* (RYMV), *Rice hoja blanca virus*, and *Rice giallume virus* have been identified as potential viruses to cause severe damage and affect the yield of rice (Hibino 1990). However, various genes have been linked to resistance against rice-infecting viruses (Wang et al. 2014b; Lee et al. 2010; Orjuela et al. 2013). Identification of SNPs linked with RTSV resistance from the genome sequence of selected rice genomes revealed that about 5% of the accessions contained the resistance alleles of *tsv1*. Moreover, RTSV resistance was a recessive trait which was tightly controlled by *tsv1* present at upstream of sequences encoding a translation initiation factor 4 (Lee et al. 2010).

Although QTLs linked to *Rice stripe virus* (RSV) resistance have already been identified (Wang et al. 2014b), available sequences may assist in identifying the structural difference within RSV resistance allele such as *STV11* (LOC_Os11g30910; encoding a sulphotransferase). It has been revealed that a 6 nucleotide deletion in *STV11* was predominantly associated with RSV resistance traits, and out of 300 accessions screened, only 0.3% accessions contained the deletion (Leung et al. 2015). Similarly, *Rice yellow mottle virus* resistance has been linked with the alleles of two genes of rice, i.e., *RYMV1* (encoding an isoform of translation initiation factor 4 gamma) and *RYMV2* (a homolog of constitutive expression of PR genes 5; Orjuela et al. 2013). It was revealed that both the alleles of *RYMV1* and *RYMV2* were present in all the accessions of *O. glaberrima*, except *RYMV1* in cultivar "Gigante." Moreover, datasets of 3000 rice accessions suggested that the RYMV resistance is an exceptionally rare trait in *O. sativa* (Leung et al. 2015). Hence, identification of genes and SNPs associated within these datasets of genome can serve as a potential resource for genomics-assisted breeding to facilitate the transfer of desired genes along with the rapid improvement of elite lines with introduced traits.

Application of GAB is so far limited in wheat. However, GAB will be an essential factor in identifying the SNPs linked with virus resistance in wheat. For example, *Wheat yellow mosaic virus* (WYMV) is one of the most important wheat infecting bymoviruses (Namba et al. 1998; Xiaoyun et al. 1998), and QTLs associated to WYMV resistance, i.e., *Qym1* and *Qym2*, (Suzuki et al. 2015) along with *QYm.njau-5A.1* (Zhu et al. 2012) and *Q. ymym* (Kojima et al. 2015) have been identified. Availability of genome sequences will assist further in identifying the genes and associated SNPs with virus resistance in wheat and support the genomics-driven breeding for genetic improvement.

5.6 Current Challenges and Future Prospects

Earlier, crop breeding practice was relying typically on the data available from phenotypic studies under field conditions, and, later, with the advances in genomics studies, the concept of molecular markers and QTLs came into the context. Modern plant breeding program integrates data from all other branches of science including genomics, proteomics, metabolomics, computational biology, and statistics to design elite varieties that have relatively high yield under adverse climatic conditions in short period of time. Availability of current genomics technologies, resources, and high-quality reference genome sequences for most major crops has led to the application of whole-genome strategies in plant breeding program. High-throughput genotyping platforms and genome-wide selections have reduced the cost and duration of selection and increase the efficiency of improvement.

It is challenging for the scientists to achieve high yield in susceptible Poaceae crops during stress conditions. To circumvent this, advanced genomics has facilitated the identification of molecular markers, QTLs, and candidate genes and construction of molecular maps throughout the genomic region of tolerant as well as susceptible varieties. MAS utilizes these resources in breeding practices for introgression of loci responsible for drought resistance from wild cultivar to susceptible high-yielding variety. MABC is the most commonly used form of MAS, but it has a limitation that it has not been very effective for developing superior lines for complex traits which are regulated by multiple loci such as drought tolerance. MARS and GS have the ability to overcome this limitation by transferring multiple genomic regions from donor to recurrent parents without affecting the characteristics of recurrent parent. But the use of MARS and GS has restricted to only multinational companies, while in public sector there are only a few examples to cite. The use of genome-wide selection in breeding needs to be extended to public sector institutions to release drought-tolerant cultivars in other graminaceous species also. In major C_4 grain crops including maize, sorghum, and foxtail millet, high-quality genome sequence is available, and many drought resistance candidate genes, molecular markers flanking these genes, and QTLs associated with this trait have also been identified. More research needs to be done for investigation and identification of complex genetic network and QTL associated with the yield-related traits such as grain size, number of tiller, size, and number of panicle. The genetic loci or QTLs associated with these traits may further transfer from high-yielding cultivar to the variety that produce limited yield but are having excellent tolerance mechanism to various abiotic stress or resistance to other biotic stress.

Acknowledgments The authors' work on cereal genetics and genomics is supported by the core grant of National Institute of Plant Genome Research, New Delhi, India. Roshan K Singh acknowledges the research fellowship received from Council of Scientific and Industrial Research, Govt. of India, India. Mehanathan Muthamilarasan and Annvi Dhaka acknowledge the research fellowship received from University Grants Commission, Govt. of India, India.

References

Alexandrov N, Tai S, Wang W, Mansueto L, Palis K, Fuentes RR, Ulat VJ, Chebotarov D, Zhang G, Li Z, Mauleon R, Hamilton RS, McNally KL (2015) SNP–seek database of SNPs derived from 3000 rice genomes. Nucleic Acids Res 43:D1023–D1027

Almeida GD, Makumbi D, Magorokosho C, Nair S, Borém A, Ribaut JM, Bänziger M, Prasanna BM, Crossa J, Babu R (2013) QTL mapping in three tropical maize populations reveals a set of constitutive and adaptive genomic regions for drought tolerance. Theor Appl Genet 126:583–600

Appleby N, Edwards D, Batley J (2009) New technologies for ultra–high throughput genotyping in plants. Methods Mol Biol 513:19–39

Babu CR, Nguyen BD, Chamarerk V (2003) Genetic analysis of drought resistance in rice by molecular markers: association between secondary traits and field performance. Crop Sci 43:1457–1469

Barnabas B, Jäger K, Feher A (2008) The effect of drought and heat stress on reproductive processes in cereals. Plant Cell Environ 31:11–38

Bedada G, Westerbergh A, Müller T, Galkin E, Bdolach E, Moshelion M, Fridman E, Schmid KJ (2014) Transcriptome sequencing of two wild barley (*Hordeum spontaneum* L.) ecotypes differentially adapted to drought stress reveals ecotype–specific transcripts. BMC Genomics 15:995

Bernardo R (2008) Molecular markers and selection for complex traits in plants: learning from the last 20 years. Crop Sci 48:1649–1664

Bernardo R, Yu J (2007) Prospects for genomewide selection for quantitative traits in maize. Crop Sci 47:1082–1090

Cardoso–Silva CB, Costa EA, Mancini MC, Balsalobre TW, Canesin LE, Pinto LR, Carneiro MS, Garcia AA, de Souza AP, Vicentini R (2014) *De novo* assembly and transcriptome analysis of contrasting sugarcane varieties. PLoS One 9:e88462

Chen G, Wang X, Hao J, Yan J, Ding J (2015) Genome–wide association implicates candidate genes conferring resistance to maize rough dwarf disease in maize. PLoS One 10:e0142001

Close TJ, Bhat PR, Lonardi S, Wu Y, Rostoks N, Ramsay L, Druka A, Stein N, Svensson JT, Wanamaker S, Bozdag S, Roose ML, Moscou MJ, Chao S, Varshney RK, Szucs P, Sato K, Hayes PM, Matthews DE, Kleinhofs A, Muehlbauer GJ, DeYoung J, Marshall DF, Madishetty K, Fenton RD, Condamine P, Graner A, Waugh R (2009) Development and implementation of high–throughput SNP genotyping in barley. BMC Genomics 10:582

Crosbie TM, Eathington SR, Johnson GR, Edwards M, Reiter R, Stark S, Mohanty RG, Oyervides M, Buehler RE, Walker AK, Dobert R, Delannay X, Pershing JC, Hall MA, Lamkey KR (2006) Plant breeding: past, present, and future. In: Lamkey KR, Lee M (eds) Plant breeding: the Arnel R. Hallauer International Symposium. Blackwell, Ames, pp 3–50

Cui Y, Lee MY, Huo N, Bragg J, Yan L, Yuan C, Li C, Holditch SJ, Xie J, Luo MC, Li D, Yu J, Martin J, Schackwitz W, Gu YQ, Vogel JP, Jackson AO, Liu Z, Garvin DF (2012) Fine mapping of the Bsr1 barley stripe mosaic virus resistance gene in the model grass Brachypodium distachyon. PLoS ONE 7, e38333

Diab AA, Teulat–Merah B, This D, Ozturk NZ, Benscher D, Sorrells ME (2004) Identification of drought–inducible genes and differentially expressed sequence tags in barley. Theor Appl Genet 109:1417–1425

Eathington SR, Crosbie TM, Edwards MD, Reiter RS, Bull JK (2007) Molecular markers in a commercial breeding program. Crop Sci 47:S154–S163

Fitzgerald TL, Powell JJ, Schneebeli K, Hsia MM, Gardiner DM, Bragg JN, McIntyre CL, Manners JM, Ayliffe M, Watt M, Vogel JP, Henry RJ, Kazan K (2015) Brachypodium as an emerging model for cereal–pathogen interactions. Ann Bot 115:717–731

Fox SE, Preece J, Kimbrel JA, Marchini GL, Sage A, Youens-Clark K, Cruzan MB, Jaiswal P (2013) Sequencing and de novo transcriptome assembly of *Brachypodium sylvaticum* (Poaceae). Appl Plant Sci 5:3

Ganal MW, Roder MS (2007) Microsatellite and SNP markers in wheat breeding. In: Varshney RK, Tuberosa R (eds) Genomic assisted crop improvement: genomics applications in crops, vol 2. Springer, Dordrecht, pp 1–24

Guo J, Su G, Zhang J, Wang G (2008a) Genetic analysis and QTL mapping of maize yield and associate agronomic traits under semi–arid land condition. Afr J Biotechnol 7:1829–1838

Guo P, Baum M, Varshney R, Graner A, Grando S, Ceccarelli S (2008b) QTLs for chlorophyll and chlorophyll fluorescence parameters in barley under post–flowering drought. Euphytica 163:203–214

Guo P, Baum M, Grando S, Ceccarelli S, Bai G, Li R, von Korff M, Varshney RK, Graner A, Valkoun J (2009) Differentially expressed genes between drought–tolerant and drought–sensitive barley genotypes in response to drought stress during the reproductive stage. J Exp Bot 60:3531–3544

Gupta PK, Varshney RK (2000) The development and use of microsatellite markers for genetic analysis and plant breeding with emphasis on bread wheat. Euphytica 113:163–185

Habash DZ, Kehel Z, Nachit M (2009) Genomic approaches for designing durum wheat ready for climate change with a focus on drought. J Exp Bot 60:2805–2815

Habier D, Fernando RL, Dekkers JC (2007) The impact of genetic relationship information on genome–assisted breeding values. Genetics 177:2389–2397

Han B, Wang C, Tang Z, Ren Y, Li Y, Zhang D, Dong Y, Zhao X (2015) Genome–wide analysis of microsatellite markers based on sequenced database in Chinese spring wheat (*Triticum aestivum* L.). PLoS One 10:e0141540

Hayano-Kanashiro C, Calderón-Vázquez C, Ibarra-Laclette E, Herrera-Estrella L, Simpson J (2009) Analysis of gene expression and physiological responses in three Mexican maize landraces under drought stress and recovery irrigation. PLoS One 4:e7531

Hibino H (1990) Resistances in rice to tungro–associated viruses. Plant Dis 74:923

Horn F, Habekuß A, Stich B (2014) Genes involved in barley yellow dwarf virus resistance of maize. Theor Appl Genet 127:2575–2584

Hospital F, Charcosset A (1997) Marker–assisted introgression of quantitative trait loci. Genetics 147:1469–1485

Huang L, Zhang F, Zhang F, Wang W, Zhou Y, Fu B, Li Z (2014) Comparative transcriptome sequencing of tolerant rice introgression line and its parents in response to drought stress. BMC Genomics 15:1026

Humbert S, Subedi S, Cohn J, Zeng B, Bi YM, Chen X, Zhu T, McNicholas PD, Rothstein SJ (2013) Genome–wide expression profiling of maize in response to individual and combined water and nitrogen stresses. BMC Genomics 14:3

Jhonson GR (2004) Marker–assisted selection in Janicke J, ed. Plant Breeding Rev 24:293–310

Johnson SM, Lim FL, Finkler A, Fromm H, Slabas AR, Knight MR (2014) Transcriptomic analysis of *Sorghum bicolor* responding to combined heat and drought stress. BMC Genomics 15:456

Jones MW, Redinbaugh MG, Anderson RJ, Louie R (2004) Identification of quantitative trait loci controlling resistance to maize chlorotic dwarf virus. Theor Appl Genet 110: 48–57

Jones MW, Boyd EC, Redinbaugh MG (2011) Responses of maize (*Zea mays* L.) near isogenic lines carrying Wsm1, Wsm2, and Wsm3 to three viruses in the Potyviridae. Theor Appl Genet 123:729–740

Kato Y, Hirotsu S, Nemoto K, Yamagishi J (2008) Identification of QTLs controlling rice drought tolerance at seedling stage in hydroponic culture. Euphytica 160:423–430

Kojima H, Nishio Z, Kobayashi F, Saito M, Sasaya T, Kiribuchi-Otobe C, Seki M, Oda S, Nakamura T (2015) Identification and validation of a quantitative trait locus associated with wheat yellow mosaic virus pathotype I resistance in a Japanese wheat variety. Plant Breeding 134:373–378

Kumar S, Sehgal SK, Kumar U, Prasad PV, Joshi AK, Gill BS (2012) Genomic characterization of drought tolerance–related traits in spring wheat. Euphytica 186:265–276

Kumpatla SP, Buyyarapu R, Abdurakhmonov IY and Mammadov JA (2012) In: IY Ibrokhim (ed) Genomics–assisted plant breeding in the 21st century: technological advances and progress. Plant Breeding, ISBN: 978–953–307–932–5, InTech, Rijeka

Lanceras JC, Pantuwan GP, Jongdee B, Toojinda T (2004) Quantitative trait loci associated with drought tolerance at reproductive stage in rice. Plant Physiol 135:384–399

Lee JH, Muhsin M, Atienza GA, Kwak DY, Kim SM, De Leon TB, Angeles ER, Coloquio E, Kondoh H, Satoh K, Cabunagan RC, Cabauatan PQ, Kikuchi S, Leung H, Choi IR (2010) Single nucleotide polymorphisms in a gene for translation initiation factor (eIF4G) of rice (*Oryza sativa*) associated with resistance to *Rice tungro spherical virus*. Mol Plant-Microbe Interact 23:29–38

Leung H, Raghavan C, Zhou B, Oliva R, Choi IR, Lacorte V, Jubay ML, Cruz CV, Gregorio G, Singh RK, Ulat VJ, Borja FN, Mauleon R, Alexandrov NN, McNally KL, Sackville HR (2015) Allele mining and enhanced genetic recombination for rice breeding. Rice 8:34

Li JY, Wang J, Zeigler RS (2014) The 3,000 rice genomes project: new opportunities and challenges for future rice research. Gigascience 3:8

Liu J, Li J, Qu J, Yan S (2015a) Development of genome–wide insertion and deletion polymorphism markers from next–generation sequencing data in rice. Rice 8:63

Liu Z, Xin M, Qin J, Peng H, Ni Z, Yao Y, Sun Q (2015b) Temporal transcriptome profiling reveals expression partitioning of homeologous genes contributing to heat and drought acclimation in wheat (*Triticum aestivum* L.). BMC Plant Biol 15:152

Luo M, Liu J, Lee RD, Scully BT, Guo B (2010) Monitoring the expression of maize genes in developing kernels under drought stress using oligo–microarray. J Integr Plant Biol 52:1059–1074

Lüpken T, Stein N, Perovic D, Habekuss A, Krämer I, Hähnel U, Steuernagel B, Scholz U, Zhou R, Ariyadasa R, Taudien S, Platzer M, Martis M, Mayer K, Friedt W, Ordon F (2013) Genomics–based high–resolution mapping of the BaMMV/BaYMV resistance gene rym11 in barley (*Hordeum vulgare* L.). Theor Appl Genet 126:1201–1212

Maccaferri M, Cane' MA, Sanguineti MC, Salvi S, Colalongo MC, Massi A, Clarke F, Knox R, Pozniak CJ, Clarke JM, Fahima T, Dubcovsky J, Xu S, Ammar K, Karsai I, Vida G, Tuberosa R (2014) A consensus framework map of durum wheat (*Triticum durum* Desf.) suitable for linkage disequilibrium analysis and genome–wide association mapping. BMC Genomics 15:873

Mace ES, Singh V, Van Oosterom EJ, Hammer GL, Hunt CH, Jordan DR (2012) QTL for nodal root angle in sorghum (*Sorghum bicolor* L. Moench) co–locate with QTL for traits associated with drought adaptation. Theor Appl Genet 124:97–109

Mammadov J, Aggarwal R, Buyyarapu R, Kumpatla S (2012) SNP markers and their impact on plant breeding. Int J Plant Genomics 2012:728398

Marconi TG, Costa EA, Miranda HR, Mancini MC, Cardoso-Silva CB, Oliveira KM, Pinto LR, Mollinari M, Garcia AA, Souza AP (2011) Functional markers for gene mapping and genetic diversity studies in sugarcane. BMC Res Notes 4:264

Mathews KL, Malosetti M, Chapman S, McIntyre L, Reynolds M, Shorter R, van Eeuwijk F (2008) Multi–environment QTL mixed models for drought stress adaptation in wheat. Theor Appl Genet 117:1077–1109

McMullen MD, Simcox KD (1995) Genomic organization of disease and insect resistance genes in maize. Mol Plant-Microbe Interact 8:811–815

Messmer R, Fracheboud Y, Bänziger M, Vargas M, Stamp P, Ribaut JM (2009) Drought stress and tropical maize: QTL–by–environment interactions and stability of QTLs across environments for yield components and secondary traits. Theor Appl Genet 119:913–930

Meuwissen TH, Hayes BJ, Goddard ME (2001) Prediction of total genetic value using genome–wide dense marker maps. Genetics 157:1819–1829

Mir RR, Zaman-Allah M, Sreenivasulu N, Trethowan R, Varshney RK (2010) Integrated genomics, physiology and breeding approaches for improving drought tolerance in crops. Theor Appl Genet 125:625–645

Muthamilarasan M, Theriappan P, Prasad M (2013) Recent advances in crop genomics for ensuring food security. Curr Sci 105:155–158

Myles S, Peiffer J, Brown PJ, Ersoz ES, Zhang Z, Costich DE, Buckler ES (2009) Association mapping: critical considerations shift from genotyping to experimental design. Plant Cell 21:2194–2202

Nagaraja RR, Madhusudhana R, Murali Mohan S, Chakravarthi DV, Mehtre SP, Seetharama N, Patil JV (2013) Mapping QTL for grain yield and other agronomic traits in post–rainy sorghum [*Sorghum bicolor* (L.) Moench]. Theor Appl Genet 126:1921–1939

Nakaya A, Isobe SN (2012) Will genomic selection be a practical method for plant breeding? Ann Bot 110:1303–1316

Namba S, Kashiwazaki S, Lu X, Tamura M, Tsuchizaki T (1998) Complete nucleotide sequence of wheat yellow mosaic bymovirus genomic RNAs. Arch Virol 143:631–643

Nelson JC, Wang S, Wu Y, Li X, Antony G, White FF, Yu J (2011) Single–nucleotide polymorphism discovery by high–throughput sequencing in sorghum. BMC Genomics 12:352

Ngugi K, Kimani W, Kiambi D, Mutitu EW (2013) Improving drought tolerance in *Sorghum bicolor* L. Moench: marker–assisted transfer of the stay–green Quantitative Trait Loci (QTL) from a characterized donor source into a local farmer variety. Int J Sci Res Knowl 1:154–162

Orjuela J, Deless EF, Kolade O, Chéron S, Ghesquière A, Albar L (2013) A recessive resistance to rice yellow mottle virus is associated with a rice homolog of the CPR5 gene, a regulator of active defense mechanisms. Mol Plant-Microbe Interact 26:1455–1463

Pandey G, Misra G, Kumari K, Gupta S, Parida SK, Chattopadhyay D, Prasad M (2013) Genome–wide development and use of microsatellite markers for large–scale genotyping applications in foxtail millet [*Setaria italica* (L.)]. DNA Res 20:197–207

Parida SK, Dalal V, Singh AK, Singh NK, Mohapatra T (2009) Genic non–coding microsatellites in the rice genome: characterization, marker design and use in assessing genetic and evolutionary relationships among domesticated groups. BMC Genomics 10:140

Peleg Z, Fahima T, Krugman T, Abbo S, Yakir D, Korol AB, Saranga Y (2009) Genomic dissection of drought resistance in durum wheat 9 wild emmer wheat recombinant inbreed line population. Plant Cell Environ 32:758–779

Petty ITD, Donald RGK, Jackson AO (1994) Multiple genetic determinants of barley stripe mosaic virus influence lesion phenotype on *Chenopodium amaranticolor*. Virology 198:218–226

Pinto RS, Reynolds MP, Mathews KL, McIntyre CL, Olivares–Villegas JJ, Chapman SC (2010) Heat and drought adaptive QTL in a wheat population designed to minimize confounding agronomic effects. Theor Appl Genet 121:1001–1021

Price AH, Townend J, Jones MP, Audebert A, Courtois B (2002) Mapping QTLs associated with drought avoidance in upland rice grown in the Philippines and West Africa. Plant Mol Biol 48:683–695

Priest HD, Fox SE, Rowley ER, Murray JR, Michael TP, Mockler TC (2014) Analysis of global gene expression in *Brachypodium distachyon* reveals extensive network plasticity in response to abiotic stress. PLoS One 9:e87499

Prince SJ, Beena R, Gomez SM, Senthivel S, Babu RC (2015) Mapping consistent rice (*Oryza sativa* L.) Yield QTLs under drought stress in target rainfed environments. Rice 8:53

Qu J, Liu J (2013) A genome–wide analysis of simple sequence repeats in maize and the development of polymorphism markers from next–generation sequence data. BMC Res Notes 6:403

Rahman H, Pekic S, Lazic–Jancic V, Quarrie SA, Shah SM, Pervez A, Shah MM (2011) Molecular mapping of quantitative trait loci for drought tolerance in maize plants. Genet Mol Res 10:889–901

Ravi K, Vadez V, Isobe S, Mir RR, Guo Y, Nigam SN, Gowda MV, Radhakrishnan T, Bertioli DJ, Knapp SJ, Varshney RK (2011) Identification of several small main–effect QTLs and a large number of epistatic QTLs for drought tolerance related traits in groundnut (*Arachis hypogaea* L.). Theor Appl Genet 122:1119–1132

Ribaut JM, Ragot M (2007) Marker–assisted selection to improve drought adaptation in maize: the backcross approach, perspectives, limitations, and alternatives. J Exp Bot 58:351–360

Ribaut JM, de Vicente MC, Delannay X (2010) Molecular breeding in developing countries: challenges and perspectives. Curr Opin Plant Biol 13:213–218

Sabadin PK, Malosetti M, Boer MP, Tardin FD, Santos FG, Guimarães CT, Gomide RL, Andrade CL, Albuquerque PE, Caniato FF, Mollinari M, Margarido GR, Oliveira BF, Schaffert RE, Garcia AA, van Eeuwijk FA, Magalhaes JV (2012) Studying the genetic basis of drought tolerance in sorghum by managed stress trials and adjustments for phenological and plant height differences. Theor Appl Genet 124:1389–1402

Sehgal D, Yadav R (2010) Molecular markers based approaches for drought tolerance. In: Jain SM, Brar DS (eds) Molecular techniques in crop improvement. Springer, New York, pp 207–230

Semagn K, Beyene Y, Warburton ML, Tarekegne A, Mugo S, Meisel B, Sehabiague P, Prasanna BM (2013) Meta–analyses of QTL for grain yield and anthesis silking interval in 18 maize populations evaluated under water–stressed and well–watered environments. BMC Genomics 14:313

Serraj R, Hash CT, Rivzi SMH (2005) Recent advances in markerassisted selection for drought tolerance in pearl millet. Plant Prod Sci 8:334–337

Soderlund C, Descour A, Kudrna D, Bomhoff M, Boyd L, Currie J, Angelova A, Collura K, Wissotski M, Ashley E, Morrow D, Fernandes J, Walbot V, Yu Y (2009) Sequencing, mapping, and analysis of 27,455 maize full–length cDNAs. PLoS Genet 5:e1000740

Sonah H, Deshmukh RK, Sharma A, Singh VP, Gupta DK, Gacche RN, Rana JC, Singh NK, Sharma TR (2011) Genome–wide distribution and organization of microsatellites in plants: an insight into marker development in Brachypodium. PLoS One 6:e21298

Suji KK, Prince KSJ, Mankhar PS, Kanagaraj P, Poornima R, Amutha K, Kavitha S, Biji KR, Gomez M, Babu RC (2012) Evaluation of rice (*Oryza sativa* L.) near isogenic lines with root QTLs for plant production and root traits in rainfed target populations of environment. Field Crop Res 137:89–96

Suzuki T, Murai MN, Hayashi T, Nasuda S, Yoshimura Y, Komatsuda T (2015) Resistance to wheat yellow mosaic virus in Madsen wheat is controlled by two major complementary QTLs. Theor Appl Genet 128:1569–1578

Tanksley SD, Nelson JC (1996) Advanced backcross QTL analysis: a method for the simultaneous discovery and transfer of valuable QTLs from unadapted germplasm into elite breeding lines. Theor Appl Genet 92:191–203

Tao Y, Liu Q, Wang H, Zhang Y, Huang X, Wang B, Lai J, Ye J, Liu B, Xu M (2013) Identification and fine–mapping of a QTL, qMrdd1, that confers recessive resistance to maize rough dwarf disease. BMC Plant Biol 13:145

The International Brachypodium Initiative (2010) Genome sequencing and analysis of the model grass *Brachypodium distachyon*. Nature 463:763–768

Tuberosa R, Salvi S (2006) Genomics–based approaches to improve drought tolerance of crops. Trends Plant Sci 11:405–412

Vargas M, van Eeuwijk FA, Crossa J, Ribaut JM (2006) Mapping QTLs and QTL x environment interaction for CIMMYT maize drought stress program using factorial regression and partial least squares methods. Theor Appl Genet 112:1009–1023

Varshney RK, Graner A, Sorrells ME (2005) Genomics–assisted breeding for crop improvement. Trends Plant Sci 10:621–630

Varshney RK, Hoisington DA, Tyagi AK (2006) Advances in cereal genomics and applications in crop breeding. Trends Biotechnol 24:490–499

Varshney RK, Langridge P, Graner A (2007) Application of genomics to molecular breeding of wheat and barley. Adv Genet 58:121–155

Varshney RK, Mohan SM, Gaur PM, Gangarao NVPR, Pandey MK, Bohra A, Sawargaonkar SL, Chitikineni A, Kimurto PK, Janila P, Saxena KB, Fikre A, Sharma M, Rathore A, Pratap A, Tripathi S, Datta S, Chaturvedi SK, Mallikarjuna N, Anuradha G, Babbar A, Choudhary AK, Mhase MB, Bharadwaj C, Mannur DM, Harer PN, Guo B, Liang X, Nadarajan N, Gowda CLL (2013) Achievements and prospects of genomics–assisted breeding in three legume crops of the semi–arid tropics. Biotechnol Adv 31:1120–1134

Varshney RK, Kudapa HB, Pazhamala L, Chitikineni A, Thudi M, Bohra A, Gaur PM, Janila P, Fikre A, Kimurto PK, Ellis NTH (2015) Translational genomics in agriculture: some examples in grain legumes. Crit Rev Plant Sci 34:169–194

Venuprasad R, Dalid CO, Del Valle M, Zhao D, Espiritu M, Sta Cruz MT, Amante M, Kumar A, Atlin GN (2009) Identification and characterization of large–effect quantitative trait loci for grain yield under lowland drought stress in rice using bulk–segregant analysis. Theor Appl Genet 120:177–190

von Korff M, Grando S, Del Greco A, This D, Baum M, Ceccarelli S (2008) Quantitative trait loci associated with adaptation to Mediterranean dry land conditions in barley. Theor Appl Genet 117:653–669

Wang D, Pan Y, Zhao X, Zhu L, Fu B, Li Z (2011) Genome–wide temporal–spatial gene expression profiling of drought responsiveness in rice. BMC Genomics 12:149

Wang S, Wong D, Forrest K, Allen A, Chao S, Huang BE, Maccaferri M, Salvi S, Milner SG, Cattivelli L, Mastrangelo AM, Whan A, Stephen S, Barker G, Wieseke R, Plieske J, International Wheat Genome Sequencing Consortium, Lillemo M, Mather D, Appels R, Dolferus R, Brown–Guedira G, Korol A, Akhunova AR, Feuillet C, Salse J, Morgante M, Pozniak C, Luo MC, Dvorak J, Morell M, Dubcovsky J, Ganal M, Tuberosa R, Lawley C, Mikoulitch I, Cavanagh C, Edwards KJ, Hayden M, Akhunov E (2014a) Characterization of polyploid wheat genomic diversity using a high–density 90,000 single nucleotide polymorphism array. Plant Biotechnol J 12:787–796

Wang Q, Liu Y, He J, Zheng X, Hu J, Liu Y, Dai H, Zhang Y, Wang B, Wu W, Gao H, Zhang Y, Tao X, Deng H, Yuan D, Jiang L, Zhang X, Guo X, Cheng X, Wu C, Wang H, Yuan L, Wan J (2014b) *STV11* encodes a sulphotransferase and confers durable resistance to rice stripe virus. Nat Commun 5:4768

Xia X, Melchinger AE, Kuntze L, Lübberstedt T (1999) Quantitative trait loci mapping of resistance to sugarcane mosaic virus in maize. Phytopathology 89:660–667

Xiao YN, Li XH, George ML, Li MS, Zhang SH, Zheng YL (2005) Quantitative trait locus analysis of drought tolerance and yield in maize in China. Plant Mol Biol Report 23:155–165

Xiaoyun L, Kashiwazaki S, Tamura M, Namba S (1998) The 3' terminal sequence of RNA1 of Wheat spindle streak mosaic virus canadian isolate (WSSMV-C). Eur J Plant Pathol 104:765–768

Xu Y, Lu Y, Xie C, Gao S, Wan J, Prasanna BM (2012) Whole–genome strategies for marker–assisted plant breeding. Mol Breed 29:833–854

Xu J, Liu L, Xu Y, Chen C, Rong T, Ali F, Zhou S, Wu F, Liu Y, Wang J, Cao M, Lu Y (2013a) Development and characterization of simple sequence repeat markers providing genome–wide coverage and high resolution in maize. DNA Res 20:497–509

Xu J, Li Y, Ma X, Ding J, Wang K, Wang S, Tian Y, Zhang H, Zhu XG (2013b) Whole transcriptome analysis using next–generation sequencing of model species *Setaria viridis* to support C4 photosynthesis research. Plant Mol Biol 83:77–87

Xu J, Yuan Y, Xu Y, Zhang G, Guo X, Wu F, Wang Q, Rong T, Pan G, Cao M, Tang Q, Gao S, Liu Y, Wang J, Lan H, Lu Y (2014) Identification of candidate genes for drought tolerance by whole–genome re–sequencing in maize. BMC Plant Biol 14:83

Xue GP, McIntyre CL, Chapman S, Bower NI, Way H, Reverter A, Clarke B, Shorter R (2006) Differential gene expression of wheat progeny with contrasting levels of transpiration efficiency. Plant Mol Biol 61:863–881

Yadav RS, Sehgal D, Vadez V (2011) Using genetic mapping and genomics approaches in understanding and improving drought tolerance in pearl millet. J Exp Bot 62:397–408

Yang P, Habekuß A, Ordon F, Stein N (2014) Analysis of bymovirus resistance genes on proximal barley chromosome 4HL provides the basis for precision breeding for BaMMV/BaYMV resistance. Theor Appl Genet 127:1625–1634

Zambrano JL, Jones MW, Brenner E, Francis DM, Tomas A, Redinbaugh MG (2014) Genetic analysis of resistance to six virus diseases in a multiple virus–resistant maize inbred line. Theor Appl Genet 127:867–880

Zhang SH, Li XH, Wang ZH, George ML, Jeffers D, Wang F, Liu XD, Li MS, Yuan LX (2003) QTL mapping for resistance to SCMV in Chinese maize germplasm. Maydica 48:307–312

Zhang Z, Deng Y, Tan J, Hu S, Yu J, Xue Q (2007) A genome–wide microsatellite polymorphism database for the indica and japonica rice. DNA Res 14:37–45

Zhou G, Xu D, Xu D, Zhang M (2013) Southern rice black–streaked dwarf virus: a white–backed planthopper–transmitted fijivirus threatening rice production in Asia. Front Microbiol 4:270

Zhou G, Zhang Q, Tan C, Zhang XQ, Li C (2015) Development of genome–wide InDel markers and their integration with SSR, DArT and SNP markers in single barley map. BMC Genomics 16:804

Zhu X, Wang H, Guo J, Wu Z, Cao A, Bie T, Nie M, You FM, Cheng Z, Xiao J, Liu Y, Cheng S, Chen P, Wang X (2012) Mapping and validation of quantitative trait loci associated with wheat yellow mosaic bymovirus resistance in bread wheat. Theor Appl Genet 124:177–188

Ziyomo C, Bernardo R (2013) Drought tolerance in maize: Indirect selection through secondary traits versus genomewide selection. Crop Sci 53:1269–1275

6 Plant Tolerance to Combined Stress: An Overview

Wusirika Ramakrishna and Anuradha Kumari

Abstract

The demand for food is predicted to increase by 70% in 2050 due to increasing world population. Efforts are being made to increase food production. However, abiotic and biotic stresses, which tend to reduce crop yield and grain quality, are hindering these efforts. Significant improvement in crop productivity can be accomplished by developing plants tolerant to multiple abiotic and biotic stresses. Plants adapt and tolerate multiple stresses using sophisticated biochemical and molecular mechanisms. These are mediated by reactive oxygen species (ROS) and phytohormones such as abscisic acid, ethylene, jasmonic acid and salicylic acid which in turn regulate ion channels and kinase cascades. Several transcription factors (TFs) including WRKY, ERF, NAC, and MYB TFs are involved in this process. Understanding these known and novel mechanisms is an important step toward developing tolerance to multiple stresses. Future directions in this field for enhancing crop productivity are discussed.

Keywords

Biotic stress • Abiotic stress • Phytohormones • Multiple stresses • Cross-tolerance

W. Ramakrishna (✉) • A. Kumari
Centre for Biochemistry and Microbial Sciences, Central University of Punjab, Bathinda, Punjab, India
e-mail: wusirika@gmail.com; wusirikark@cup.ac.in

M. Senthil-Kumar (eds.), *Plant Tolerance to Individual and Concurrent Stresses*,
DOI 10.1007/978-81-322-3706-8_6

6.1 Introduction

Every year, there is a potential decrease in crop yield due to biotic and abiotic stresses. Plants encounter abiotic stresses, which include drought, soil salinity, heavy metal contamination, mineral deficiency, and high and low temperatures, under field conditions. Biotic stresses such as infectious bacteria, fungi, viruses, and nematodes also affect crop productivity. Abiotic stress has very high impact on the growth of crops and, therefore, responsible for severe losses. Stress-induced biochemical and physiological alterations in plants are the result of abiotic and biotic stresses or adaptation to tolerate them or both. Plants adapt to various environmental stresses via activation of cascades of molecular networks involved in stress sensitivity and the expression of stress-related genes and metabolites (Vinocur and Altman 2005; Rejeb et al. 2014). To improve the production efficiency, crop varieties resistant to various abiotic and biotic stresses are being developed using biotechnological approaches.

Plants have evolved sophisticated mechanisms to ascertain external signals and respond to different environmental conditions for their survival. One of the crucial steps in plant defense is the perception of stress at the right time so that the plants respond to it in a rapid and efficient manner. After recognizing stress, the plant basal defense mechanisms activate the complex signaling cascades of defense which differ from one stress to another (AbuQamar et al. 2009; Andreasson and Ellis 2010). How the plants cope up with all stress conditions (abiotic and biotic) is an important step to elucidate the complex regulatory mechanisms and signaling pathways, which determine stress tolerance. Earlier studies reported that there is a cross talk between different stress responses and hormonal signaling with the upregulation of WRKY transcription factor (TF)-encoding genes in both bacterial and drought stresses in *Arabidopsis* and rice (Shaik and Ramakrishna 2013). Downstream targets of WRKY transcription factors include mitogen-activated protein (MAP) kinases. Drought-induced WRKY TFs in rice were predicted to interact with several other proteins suggesting a key role in stress response (Shaik and Ramakrishna 2012). In addition to WRKY TFs, ERF, NAC, and MYB TFs play an important role in regulating multiple stress tolerance (Shaik and Ramakrishna 2014; Wang et al. 2016). Transgenic plants overexpressing these TF-encoding genes and others including *DREB* genes have been shown to confer tolerance to more than one abiotic and/or biotic stress (Jan et al. 2013; Shaik and Ramakrishna 2014). The knowledge about plant responses to combined stress conditions will help in managing the growth and development of plants (Pandey et al. 2015). When plants encounter abiotic and biotic stresses, the levels of reactive oxygen species (ROS) and phytohormones such as abscisic acid, ethylene, jasmonic acid, and salicylic acid are altered, which in turn activate kinase cascades and regulate ion channels (Saxena et al. 2016). These changes in the plant machinery lead to an adequate defense system which results in an increase in plant tolerance to lower the biological damages caused by different stresses (Verma et al. 2016).

6.2 Shared Versus Unique Plant Responses to Multiple Stresses

Plant responses to stress combinations are governed by various factors such as how severe are the stresses, the age of plants, and susceptibility of plants to pathogens. The shared responses exhibit common physiological and molecular events, whereas some physiological traits are unique to individual stresses. The combination of two abiotic stresses can either have additive or antagonistic effects on each other. For example, a combination of drought and salt stress led to a severe reduction in net photosynthetic rate, stomatal conductance, and enhanced oxidative damage in *Hordeum spontaneum* (Ahmed et al. 2013). It also resulted in increased Na^+ accumulation in roots as compared to leaves and stems, while Na^+ accumulated in shoots under salinity stress only. Similarly, under heat stress, plants open their stomata by transpiration to cool their leaves, but plants would not be able to open their stomata when heat and drought stress are given together as it would lead to high temperature in leaves and causes wilting (Rizhsky et al. 2002). Further, heat stress can increase transpiration which in turn would increase salt and heavy metal uptake (Mittler and Blumwald 2010). Drought and heat stress combination enhanced overall damage when drought and heat stress were given together as compared to individual stresses because these two stresses share some common physiological traits (Pandey et al. 2015). Heat stress when combined with salt stress resulted in different response compared to individual stress. Proline is a predominant osmoprotectant which accumulates in plants under salt stress. However, when *Solanum lycopersicum* plants were treated with heat and salt stresses together, accumulation of glycine betaine and trehalose increased for the protection of membranes and photosynthetic proteins (Rivero et al. 2014). In this study, trehalose levels were eightfold in combined stress compared to control plants with no change in heat treated and lower levels in salt treated compared to control plants.

Combined biotic and abiotic stresses tend to conquer the stress adaptation strategies which are different and sometimes show contrasting results to those seen under individual stresses. Exposure of plants to multiple stresses confer cross-tolerance where one stress protects them from other stresses by utilizing regulatory systems which allow rapid adaptation to the changing environment (Bowler and Fluhr 2000; Jalmi and Sinha 2015). For example, in tomato plants wounding increased salt tolerance accompanied by the upregulation of calmodulin-like activities involved in downstream signaling required for cross-tolerance (Capiati et al. 2006). Similarly, infection by *Pseudomonas syringae* pv. tomato induces resistance to *Helicoverpa zea*, a herbivore insect (Rejeb et al. 2014). Abiotic stress along with biotic stress either increases or inhibits the effect of biotic stress that leads to enhanced or reduced susceptibility to pathogens. Thus, abiotic stress can change plant tolerance or pathogen susceptibility by different mechanisms, thereby modifying plant-pathogen interactions. Therefore, it is important to study the biochemical and molecular responses of plants under stress combinations than individual stresses to develop efficient strategies to develop plants tolerant to multiple stresses. Two independent approaches have been proposed for developing plants with tolerance to combined biotic and abiotic stress (Kissoudis et al. 2014). The first approach is to manipulate common regulators such as WRKY, MYB, and NAC transcription factors and flavonoid metabolism involved in both biotic and abiotic stress. The second approach is to pyramid different genes which confer resistance to a specific disease and abiotic stress independently without any negative effect on plant growth-related traits (Kissoudis et al. 2014). The key to any approach for developing plants with cross-tolerance to combined stress is to give specific biotic or abiotic stress in the correct magnitude so that they exhibit minimal or no effect on plant growth and development.

6.3 Role of Phytohormones in Combined Stress

Phytohormones have an important role in plant responses under biotic as well as abiotic stresses via signaling pathways. It is well known that the hormone ABA controls the abiotic stress responses, while SA, JA, and ethylene (ET) signaling pathways control the defense against biotic stress. However, recent studies revealed that ABA also controls the biotic signaling pathways either synergistically or antagonistically (Asselbergh et al. 2008; Yasuda et al. 2008; Kissoudis et al. 2014). ABA is known to have a central role in abiotic stress responses as it reduces transpiration rate by the stomatal closure (Pantin et al. 2013) and regulates root growth and ion channels (Duan et al. 2013). Also, stomatal closure induced by ABA prevents the entry of microbes through open stomata. Therefore, ABA is considered to have a central role in cross talk between biotic and abiotic stress responses, and production of ABA can be a major factor in plant response to multiple stresses (Atkinson and Urwin 2012). ABA also acts as a negative regulator of disease resistance (Koga et al. 2004; Mauch-Mani and Mauch 2005). For example, tomato mutant *sitiens*, which is ABA deficient, has increased resistance to pathogens (Thaler and Bostock 2004). However, on treatment with ABA, the susceptibility to pathogens was restored in *sitiens*. ABA regulates other plant hormones to enhance the production of antimicrobial compounds to stop the invasion of pathogens (Ton et al. 2009).

Recent studies have identified hormones such as auxins (indole-3-acetic acid [IAA]), cytokinins, and gibberellins, which are primarily involved in regulating plant growth and development, with an additional role in biotic and abiotic stress tolerance (Vleesschauwer et al. 2014). Most of these hormones at lower levels promote disease resistance and at higher levels promote plant susceptibility to pathogens with some exceptions. Lower levels of IAA enhanced susceptibility of rice to *Magnaporthe oryzae*, *Xoo*, and *Xoc*. Several pathogens can produce cytokinins or promote its production by plants resulting in enhanced pathogen virulence and suppression of host immunity. However, recent data identified opposite effect in some cases where cytokinins enhanced plant immunity (SA-independent) as observed in tobacco inoculated with *Pseudomonas syringae* pv. *tobacco*. Cytokinins cross talk primarily with ABA under abiotic stress and SA under biotic stress (O'Brien and Benková 2013). A complete understanding of the role of phytohormones in combined stress would enable us to design efficient strategies to enhance crop productivity.

6.4 Reactive Oxygen Species and Transcription Factors as Common Links Between Biotic and Abiotic Stress

Reactive oxygen species (ROS) act as secondary messengers and are required at normal levels for plant growth and development. ROS species include hydrogen peroxide, superoxide, hydroxyl radical, and singlet oxygen. Enzymes such as ascorbate peroxidase, catalase, and superoxide dismutase maintain ROS homeostasis in plants. Biotic and abiotic stress results in higher levels of ROS which act as a common signaling molecule. Higher ROS levels can interact and damage all four macromolecules. During stress, cross talk has been reported between ROS such as hydrogen peroxide and hormones, ABA, SA, JA, and ethylene (Sewelam et al. 2016; Saxena et al. 2016). It acts as a second messenger in ABA-regulated stomatal closure. ROS is a common signaling molecule for mitogen-activated protein kinase (MAPK) cascade which is shared by both biotic and abiotic stress (Jalmi and Sinha 2015). The same kinase activated by ROS gives a different response depending on the nature of the stress (biotic or abiotic). For instance, ROS regulates MKK6 which in turn regulates MPK3 and MPK6, but the final outcome of innate immunity or tolerance to cold, drought, or salt is dependent on biotic or abiotic stress. ROS regulate transcription factors through MAPK pathway. Transcription factors in turn regulate stress-responsive genes whose protein products confer stress tolerance and act as important signaling molecules as part of plant stress response. Transgenic plants overexpressing AP2/EREBP, MYB, WRKY, NAC, and bZIP transcription factor-encoding genes conferred tolerance to two or more abiotic stresses (Wang et al. 2016). However, most of these studies were conducted in greenhouse.

Genetic manipulation of genes encoding transcription factors has advantages and disadvantages. Transcription factors can regulate multiple genes not only involved in stress but also other biological processes. If we do not understand the complete repertoire of genes regulated by a specific transcription factor, its manipulation may lead to plants with undesirable effects with reference to growth and yield attributes. In addition to gene regulation, transcription factors can interact with other proteins and metabolites. It is often difficult to understand all protein-protein and protein-metabolite interactions of a transcription factor. On the contrary, accurate selection of a transcription factor which would confer tolerance to combined stress and whose mode of action is well understood would be highly beneficial for generating a transgenic plant.

6.5 Future Perspectives on Exploiting Combined Stress for Enhancing Crop Productivity

Most of the studies on plants subjected to combined stress have been carried out in greenhouse. Field trials at multiple locations and multiple years with different combinations of biotic and abiotic stress are needed to gain knowledge about their physiological effects on different plant species. This information when combined with an integrated systems biology approach (transcriptomics, proteomics, and metabolomics) would identify key players and biological pathways involved in combined stress response. Non-transgenic approaches of genome editing (Luo et al. 2015) can be used to modify key genes required to enhance tolerance to combined stress. Alternately, plant varieties or land races with upregulated or downregulated genes associated with tolerance to combined stress can be identified. Another interesting approach is the use of plant growth promoting bacteria (PGPB) to enhance tolerance to combined stress (Li and Ramakrishna 2011; Li et al. 2014; Dhawi et al. 2015). This approach would utilize naturally occurring bacteria eliminating the need for genetic modification of plants.

Several decades of research has provided a plethora of information about the diverse roles of plant hormones on plant growth and development as well as biotic and abiotic stress. However, our understanding of the cross talk among these hormones under combined stress is far from complete. Once we gain further insights into the cross talk among hormones and associated proteins and metabolites in a model crop plant like rice subjected to combined stress, the next task would be to develop plants tolerant to combined stress. This task may need regulation of several genes involved in one or more biological pathways which not only provide tolerance to combined stress but also maintain plant fitness. This can be achieved using engineered mini-chromosomes which allow transfer of multiple genes on an independent chromosome. This system has been developed for maize, rice, and other crop plants (Birchler 2015).

References

AbuQamar S, Luo H, Laluk K, Mickelbart MV, Mengiste T (2009) Crosstalk between biotic and abiotic stress responses in tomato is mediated by the AIM1 transcription factor. Plant J 58(2):347–360

Ahmed IM, Cao F, Zhang M, Chen X, Zhang G, Wu F (2013) Difference in yield and physiological features in response to drought and salinity combined stress during anthesis in Tibetan wild and cultivated barleys. PloS One 8(10):e77869

Andreasson E, Ellis B (2010) Convergence and specificity in the *Arabidopsis* MAPK nexus. Trends Plant Sci 15:106–113

Asselbergh B, De Vieesschauwer D, Hofte M (2008) Global switches and fine-tuning-ABA modulates plant pathogen defense. Mol Plant-Microbe Interact 21:709–719

Atkinson NJ, Urwin PE (2012) The interaction of plant biotic and abiotic stresses: from genes to the field. J Exp Bot 63:3523–3543. doi:10.1093/jxb/ers100

Birchler JA (2015) Engineered minichromosomes in plants. Chromosom Res 23:77–85

Bowler C, Fluhr R (2000) The role of calcium and activated oxygen as signals for controlling cross-tolerance. Trends Plant Sci 5:241–246

Capiati DA, Pais SM, TellezIñon MT (2006) Wounding increases salt tolerance in tomato plants: evidence on the participation of calmodulin-like activities in cross tolerance signaling. J Exp Bot 57:2391–2400

Dhawi F, Datta R, Ramakrishna W (2015) Mycorrhiza and PGPB modulate maize biomass, nutrient uptake and metabolic pathways in maize grown in mining-impacted soil. Plant Physiol Biochem 97:390–399

Duan L, Dietrich D, Ng CH, Chan PMY, Bhalerao R, Bennett MJ, Dinneny JR (2013) Endodermal ABA signaling promotes lateral root quiescence during salt stress in Arabidopsis seedlings. Plant Cell 25(1):324–341

Jalmi SK, Sinha AK (2015) ROS mediated MAPK signaling in abiotic and biotic stress – striking similarities and differences. Front Plant Sci 6:769

Jan AT, Singhal P, Haq QMR (2013) Plant abiotic stress: deciphering remedial strategies for emerging problem. J Plant Interact 8(2):97–108

Kissoudis C, van de Wiel C, Visser RGF, van der Linden G (2014) Enhancing crop resilience to combined abiotic and biotic stress through the dissection of physiological and molecular crosstalk. Front Plant Sci 5:207

Koga H, Dohi K, Mori M (2004) Abscisic acid and low temperatures suppress the whole plant-specific resistance reaction of rice plants to the infection of *Magnaporthe grisea*. Physiol Mol Plant Pathol 65(1):3–9

Li K, Ramakrishna W (2011) Effect of multiple metal resistant bacteria from contaminated lake sediments on metal accumulation and plant growth. J Hazard Mater 189:531e539

Li K, Pidatala VR, Shaik R, Datta R, Ramakrishna W (2014) Integrated metabolomics and proteomic approaches dissect the effect of metal-resistant bacteria on maize biomass and copper uptake. Environ Sci Technol 48:1184–1193

Luo S, Li J, Stoddard TJ, Baltes NJ, Demorest ZL, Clasen BM, Coffman A, Retterath A, Mathis L, Voytas DF, Zhang F (2015) Non-transgenic plant genome editing using purified sequence-specific nucleases. Mol Plant 8(9):1425–1427

Mauch-Mani B, Mauch F (2005) The role of abscisic acid in plant-pathogen interactions. Curr Opin Plant Biol 8:409–414

Mittler R, Blumwald E (2010) Genetic engineering for modern agriculture: challenges and perspectives. Annu Rev Plant Biol 61:443–462

O'Brien JA, Benková E (2013) Cytokinin cross-talking during biotic and abiotic stress responses. Front Plant Sci 4:451

Pandey P, Ramegowda V, Senthil-Kumar M (2015) Shared and unique responses of plants to multiple individual stresses and stress combinations: physiological and molecular mechanisms. Front Plant Sci 6:723

Pantin F, Monnet F, Jannaud D, Costa JM, Renaud J, Muller B, Bhalerao R, Simonneau T, Genty B (2013). The dual effect of abscisic acid on stomata. New Phytol 197(1): 65–72. doi:10.1111/nph.12013

Rejeb IB, Pastor V, Mauch-Mani B (2014) Plant responses to simultaneous biotic and abiotic stress: molecular mechanisms. Plants 3(4):458–475

Rivero RM, Mestre TC, Mittler R, Rubio F, Garcia-Sanchez F, Martinez V (2014) The combined effect of salinity and heat reveals a specific physiological, biochemical and molecular response in tomato plants. Plant Cell Environ 37(5):1059–1073

Rizhsky L, Liang H, Mittler R (2002) The combined effect of drought stress and heat shock on gene expression in tobacco. Plant Physiol 130(3):1143–1151

Saxena I, Srikanth S, Chen Z (2016) Cross talk between H_2O_2 and interacting signal molecules under plant stress response. Front Plant Sci 7:570

Sewelam N, Kazan K, Schenk PM (2016) Global plant stress signaling: reactive oxygen species at the cross-road. Front Plant Sci 7:187

Shaik R, Ramakrishna W (2012) Bioinformatic analysis of epigenetic and microRNA mediated regulation of drought responsive genes in rice. PloS One 7(11):e49331

Shaik R, Ramakrishna W (2013) Genes and co-expression modules common to drought and bacterial stress responses in Arabidopsis and rice. PloS One 8(10):e77261
Shaik R, Ramakrishna W (2014) Machine learning approaches distinguish multiple stress conditions using stress-responsive genes and identify candidate genes for broad resistance in rice. Plant Physiol 164:481–495
Thaler JS, Bostock RM (2004) Interactions between abscisic acid mediated responses and plant resistance to pathogens and insects. Ecology 85:48–58
Ton J, Flors V, Mauch-Mani B (2009) The multifaceted role of ABA in disease resistance. Trends Plant Sci 14(6):310–317
Verma V, Ravindran P, Kumar PP (2016) Plant hormone-mediated regulation of stress responses. BMC Plant Biol 16:86
Vinocur B, Altman A (2005) Recent advances in engineering plant tolerance to abiotic stress: achievements and limitations. Curr Opin Biotechnol 16(2):123–132
Vleesschauwer DD, Xu J, Höfte M (2014) Making sense of hormone-mediated defense networking: from rice to Arabidopsis. Front Plant Sci 5:611
Wang H, Wang H, Shao H, Tang X (2016) Recent advances in utilizing transcription factors to improve plant abiotic stress tolerance by transgenic technology. Front Plant Sci 7:67
Yasuda M, Ishikawa A, Jikumaru Y, Seki M, Umezawa T, Asami T, Nakashita AM, Kudo T, Shinozaki K, Yoshida S, Nakashita H (2008) Antagonistic interaction between systemic acquired resistance and the abscisic acid-mediated abiotic stress response in *Arabidopsis*. Plant Cell 20:1678–1692

Drought and Heat Tolerance in Chickpea: Transcriptome and Morphophysiological Changes Under Individual and Combined Stress

7

Renu Yadav, Sumandeep Juneja, Priyanka Singh, and Sanjeev Kumar

Abstract

Increase in global temperature due to climate change is the major concern and known to have detrimental effect on many agricultural crops. Chickpea (*Cicer arietinum* L.) is an important legume grown in the arid and semiarid region of the world. Chickpea being a heat sensitive crop is greatly affected by heat stress during both vegetative and reproductive stages. Stress resistance mechanism of chickpea involves signal perception, transduction, and subsequent activation of stress-responsive genes encoding reactive oxygen species (ROS) scavenging and osmolyte, chaperones, and aquaporins. There are different stress perception and signaling pathways, both in drought and high-temperature stress, but some common pathways also exist between the two mechanisms. The present chapter summarizes the cross talk between the drought and heat stress and the molecular mechanism underlying individual stress. Field plants are exposed to multiple stresses, and the combined effect might be antagonistic or synergistic. Hence, improving stress tolerance of plants requires a reevaluation, taking into account the effect of multiple stresses on plant metabolism and stress resistance.

Keywords

Chickpea • Drought stress • Heat stress • Molecular mechanism • Stress resistance • Transcription factors

R. Yadav • S. Juneja • P. Singh • S. Kumar (✉)
Centre for Plant Sciences, School of Basic and Applied Sciences, Central University of Punjab, Bathinda 151001, Punjab, India
e-mail: sanjeevpuchd@gmail.com

M. Senthil-Kumar (eds.), *Plant Tolerance to Individual and Concurrent Stresses*,
DOI 10.1007/978-81-322-3706-8_7

7.1 Introduction

Chickpea (*Cicer arietinum* L.) is an important annual cool season food legume of arid and semiarid regions of North Africa and West Asia (Saxena et al. 1996). It is a self-pollinated, diploid (2n = 16) grain legume with genome size of 732 Mb (Azimi et al. 2015). Early researchers believed that the chickpea originated in the southern Caucasus and northern Persia, but the report confirmed the origin to be southeastern Turkey. It was introduced to India from Turkey during the Bronze Age. Two annual species of chickpea (*C. echinospermum* and *C. reticulatum*) were found in southeastern part of Turkey adjoining Syria (Ladizinsky and Adler 1976).

India ranks first among the chickpea-producing countries. Gujarat, Andhra Pradesh, Uttar Pradesh, Madhya Pradesh, Maharashtra, Rajasthan, and Karnataka are the major chickpea-producing states in India which contribute 95% production (ICRISAT). Chickpea is grown in many parts of the world and yields a total of about 10 million tons from a planted area of 14 million hectares. Two varieties of chickpea, namely, Desi (colored seed coat, small seed size or "microcarpa") and Kabuli (pale-colored seed coat with large seeds, "macrocarpa"), are grown worldwide (Van der Maesen 1972). In India, Desi variety is largely produced.

7.2 Chickpea Versatility and Use

The cultivation of chickpea is of particular importance in terms of food security in the developing world. Chickpea seeds are a primary source of protein (Varshney et al. 2013). It is valued among other crops because of its high nutritive content (20–30% protein, ~40% carbohydrates, and remaining dietary fibers and other constituents) (Hulse 1989). The leaves, seeds, and pods secrete malic acid and oxalic acid. These acids lower the blood cholesterol level and are also used as medicine for several diseases like aphrodisiac, bronchitis, catamenia, cholera, constipation, diarrhea, dyspepsia, flatulence, snakebite, sunstroke, and warts (Duke 2012).

7.3 Ecological Scenario

Chickpea is mainly self-pollinated and cross-pollination (0–1%) is rare (Ellis et al. 1994; Singh et al. 1994). It is usually grown as a rainfed cool-weather crop or as a dry-climate crop in semiarid regions. Optimum conditions for chickpea growth include temperature of 18–26 °C during day, 21–29 °C during night, and an annual rainfall of 600–1000 mm (Duke 2012). It is a long-day plant, but flowering occurs in every photoperiod (Kumar et al. 2013).

7.4 Factors Affecting Chickpea Production

India accounts for the largest production of chickpea around the world, but still the percent yield was negative in the year 2012–2013. Chickpea productivity is constrained by several abiotic stresses (Singh et al. 1994; Gaur et al. 2007). Among the different abiotic stresses, drought and high temperature are most lethal for crop growth and limit chickpea yield substantially (Basu et al. 2009). The growth of chickpea takes place between 15 °C and 30 °C, and temperature above/below this range is lethal for it. During reproductive stage, crop experiences cool (5–8 °C) and frosty nights (0–18 °C) in the early vegetative stage and warm (20–27 °C) to hot (>38 °C) air temperature during the day in the late March and April (Summerfield et al. 1984). Like other winter-season legumes like lentil, peas, and fava bean, chickpea is more prone to heat stress than warm-season legumes like cowpea, soybean, groundnut, pigeon pea, and mung bean.

7.4.1 Influence of Drought Stress on Chickpea

Plants, as immobile organisms, evolved appropriate mechanisms to cope with temporary water limitations to carry out their growth and reproduction. The resistance of the plant to drought can be subdivided into the escape, avoidance, and tolerance strategies. Escape involves reproduction before the onset of stress and use of reserves for seed production. Avoidance includes the closure of stomata, reduction in leaf area, and senescence of older leaves. Lastly, tolerance involves scavenging of ROS. Plants respond to drought at the molecular, cellular, and physiological levels. This response against drought depends on the genotype (Rampino et al. 2006), the length of stress and severity of water loss (Araus et al. 2002), and the stage of development of plants (Zhu et al. 2000). Anatomical adaptation includes rolling of leaves, floral abscission, and alteration in cuticle permeability and floral induction (Bartels and Sunkar 2005; Seki et al. 2007). Drought stress executes a meaningful impact on chickpea yield throughout the world which causes a significant yield loss, which promoted chickpea research work to develop drought-tolerant cultivars.

7.4.2 Accumulation of Osmolytes and Sugars

Several sugars such as raffinose, stachyose, and trehalose and sugar alcohols like sorbitol and mannitol, amino acids like proline, amines such as glycine betaine, and polyamines such as putrescine (put), spermidine (spd), and spermine (spm) accumulate and play an important role in drought tolerance of plants. Fructans – a family of oligo- and poly-fructose – also play a significant role in drought stress (Seki et al. 2007).

7.4.3 Role of Phytohormones in Drought Stress Tolerance

Plant hormones play a central role in plant ability to adapt to abiotic stress (Santner and Estelle 2009). The role of ABA is the most studied hormone during drought stress, but the role of cytokinins, brassinosteroids, and auxins is being now researched. ABA plays the crucial role during drought stress by regulating the stomatal opening, but now other hormones like CK, ethylene, BR, JA, SA, and NO are also studied. Cross talk among different plant hormones results in synergistic and antagonistic interactions which play an important role in providing tolerance to abiotic stress (Peleg and Blumwald 2011).

7.5 Regulation at Transcriptional Level

Response to drought stress involves ABA-dependent, ABA-independent, and ubiquitination-related mechanism. The long-term signal of drought stress involves ABA. ABA signaling involves protein kinase or phosphatase cascade involving Ca^{2+}. In *Arabidopsis thaliana*, the transmembrane histidine kinase (ATHK1) acts as osmosensor by perceiving the drought stress signal. However, the changes in the cytoplasm Ca^{+} concentration integrate these pathways. The kinases cascade and then activate various transcription factors: ABA independent (DREB2, ZF-HD, DREB1/CBF) and ABA dependent (AREB/AB, MYC, MYB, and NAC) (Urano et al. 2010).

Many research papers have provided enough evidence to show that NAC transcription factors impart drought tolerance to chickpea (Nguyen et al. 2015). Very little knowledge is available about how ABA is perceived, although ABA is recognized both inside and outside but till now no receptor is identified for, but pieces of evidence suggest that it is recognized both inside and outside the cell (Bray 1997). ABA changes the internal pH and causes the depolarization of the membrane and regulates different type of ion channel. It activates S-type anion channels and the outward-rectifying K^{+} channel, also inhibits the inward-rectifying K+ channel and the plasma membrane H^{+}-ATPase (Leung and Giraudat 1998). First response of ABA in guard cell is to increase the concentration of Ca^{2+} in the cytoplasm. Stress-dependent gene expression showed Ca^{2+-}mobilizing second messenger cyclic adenosine 5′-diphosphate-ribose (cADPR) plays a major role in ABA response (Wu et al. 1997). The exact role of ubiquitination in abiotic stress response is limited. The transcription factors DREB1A/CBF3 and DREB2A specifically interact with cis-acting DRE/CRT involved in cold and drought stress-responsive gene expression in *Arabidopsis* (Alexandre et al. 2009). Two novel proteins DRIP1 and DRIP2 act as novel regulators in drought-responsive genes and target DREB2A protein to 26s proteasome proteolysis (Qin et al. 2008). DRIP1 and DRIP2 interact with DREB2A in the nucleus and function as E3 ubiquitin ligase and mediate DREB2A ubiquitination. Other system imparting drought resistance mechanism involves vacuolar membrane transport, unfolded protein response (UPR) pathway genes, and ROS signaling (Meiri and Breiman 2009).

7.5.1 Protein Kinases

ABA induces MAPK kinase activation. It is also known that MAPK, MAPKKK, and ribosomal S6 kinase are also induced by different kinds of environmental stresses. Drought and salinity stress induce the expression of CDPK (calcium-dependent protein kinase). AAPK (serine/threonine protein kinase) plays an important role in the Ca^{2+}-independent ABA signaling pathways in guard cells. Other kinases responsive to ABA are PKABA1 and RPK1 (receptor-like kinase), which are induced by various environmental stresses (Campalans et al. 1999).

7.5.2 Protein Phosphatase

Protein phosphatases (PPases) are essential components of the ABA signal transduction cascades. Serine/threonine protein phosphatases 1, 2A, 2B, and 2C are implicated in the ABA-mediated stomatal closure. Protein tyrosine phosphatase (PTPase) regulates MAPKs under environmental stress (Campalans et al. 1999).

7.5.3 Other Signaling Molecules

Under environmental stresses phospholipase C (PLC) is induced. PLC produces IP3 (inositol 1,4,5-triphosphate) and DAG (1,2-diacylglycerol). IP3 mediates the release of Ca^{2+} from internal stores into the cytoplasm in mammalian cells. In plants IP3 also acts as an intermediate in the signal transduction pathways under stress conditions. DAG is converted to phosphatidic acid (PPA). PPA is also the product of the activity of phospholipase D (PPD) and both mediate the ABA action (Campalans et al. 1999).

7.5.4 Gene Induction During Drought Stress

ABA causes induction of many genes; some are fast responsive and others slow responsive genes. The ABA-responsive element (ABRE) functions as a cis-acting element involved in ABA-regulated gene expression that does not require protein biosynthesis. The ABRE element is a defined sequence of 8–10 base pairs with an ACGT core sequence. bZIP is identified as the binding protein that responds slowly to ABA. CE1 and CE3 (coupling elements) are active in combination with ABRE but not alone. Myb and Myc are other elements to which MYB and MYC proteins bind. Drought-responsive element (DRE) (also called C-repeat) binds DREB2 and AP2 proteins.

7.6 Metabolic Level Changes

7.6.1 Protein Functioning

Proteins and membranes are stabilized by sugars such as trehalose and sucrose, and osmolytes like proline and glycine betaine play an important role during desiccation. Transport proteins like ion channels and aquaporins are involved in controlling the K^+ uptake and maintain the water status, respectively (Chrispeels and Maurel 1994). During drought stress, proteins such as heat shock proteins are induced, which are involved in protein repair and folding. HSPs which are expressed in the absence of heat shock are referred as HSP cognates. HSP70 plays an important role in protein folding and renaturation of denatured proteins during stress conditions. A low molecular weight HSP from *Phaseolus vulgaris* is induced by water deficit, ABA, and heat shock during late embryogenesis (Colmenero-Flores et al. 1997). Protease is required for the degradation of polypeptides denatured during cellular stress. The mobilized amino acids are then used for the synthesis of new proteins in response to stress or for osmotic adjustment. Drought stress produces activated oxygen species such as superoxide radicals and H_2O_2, which are managed by antioxidative activity of either nonenzymatic (Vit C and E, glutathione, flavonoids, alkaloids, carotenoids, and polyamines) or enzymatic identities (catalase, superoxide dismutase, peroxide, and metallothionein). Ferritin may play a role in protecting cells from the oxidative damage caused by stress by sequestering the intracellular iron involved in the generation of various reactive hydroxyl radicals through a Fenton reaction.

7.6.2 Cell Wall Alterations

Drought stress causes alterations in the chemical composition and physical properties of the cell wall. Proline-rich proteins (PRP) have been identified in pea, as a consequence of crosslinking between the PRP proteins and cell wall (Colmenero-Flores et al. 1997). Fatty acid metabolism-related genes may participate in the repair of stress-induced damage in membranes, to regulate permeability to toxic ions and fluidity of the membrane (Holmberg and Bülow 1998).

7.6.3 Lipid Transfer Proteins (LTPs)

LTPs deposit lipophilic compounds in the cell wall and also have the ability to transfer lipids between membrane vesicles in vitro. LTP expression can be induced by different kinds of abiotic stresses such as cold, drought, and osmotic stress. LTPs are also involved in the repair of stress-induced membrane damage and change the lipid composition in response to higher temperature and regulate their permeability to toxic ions (Kader 1997).

7.6.4 RAB/LEA/Dehydrins

LEA proteins were first described in cotton as late embryogenesis abundant proteins. They are accumulated in dry seeds and vegetative tissues during different kinds of environmental stresses, including drought, low temperatures, and high salt. LEA proteins protect cells and proteins against dehydration by maintaining protein and membrane structure, sequester ion, bind water, and also act as molecular chaperones (Close and Bray 1993). Stress-induced protein family members have been named using different criteria such as developmental characteristics and regulation of expression (responsive to ABA, RAB; induced by dehydration, DHN). Different protein sizes have been described: the smallest, 9-kDa Wsi724 from rice, and the largest 200-kDa Wcs20 from wheat. RAB/LEA/DHN proteins can be located in the nucleus and in the cytoplasm. Some dehydrins are constitutively expressed in pea and *Arabidopsis*, to provide the early protection to stress so that plant can get the enough time to induce other stress proteins (Close 1996). Recent studies have also shown relation between heat stress-induced gene expression and DREB2A gene (Sato et al. 2014).

7.7 Effect of Heat Stress on Chickpea

Due to global warming, chickpea experiences abnormally high temperature (>35 °C) during reproductive stage, which is a major constraint for its productivity (Basu et al. 2009). According to Chanders et al. (2008), an increase in the seasonal temperature of 1 °C can reduce chickpea yield by 53–300 kg/ha in different regions of India.

Depending on timing, duration, and interaction, observed heat stress can be grouped into chronic and acute. Chronic type of heat stress occurs at any stage of crop growth and results in substantial crop loss and finally to yield loss. Acute type of heat stress of relatively short duration can occur at any stage of crop growth, often leading to reduced yield (Devasirvatham et al. 2012). The adaptive strategies of chickpea plants to high-temperature stress are classified into heat escape, avoidance, and tolerance (Wery et al. 1993). To escape from heat stress, plants may undergo early flowering and maturity. The avoidance strategy includes changing in leaves orientation, transpiration, and reflectance of light (Wery et al. 1993). Heat tolerance mechanism includes alteration of membrane lipid composition, membrane stability, heat shock protein accumulation, and formation of osmolytes (e.g., proline) particularly in pollen. Chickpea adapts to a high temperature through an escape mechanism. But still during reproductive development, high temperature can cause significant yield loss (Kumar et al. 2013).

High temperature disrupts the membrane integrity by increasing the fluidity of lipid bilayer and denaturing membranous proteins. This enhances the permeability of membranes and causes increased loss of electrolytes (Maestri et al. 2002). During high temperature, the ratio of unsaturated to saturated fatty acid decreases to one-third of the levels at normal temperatures (Qu et al. 2013).

Chickpea has relatively narrow genetic base, which is another reason for the detrimental effect of high temperature on growth and reproductive physiology (Abbo et al. 2003). Heat stress negatively impacts reproductive growth by leading to impairment of micro- and megasporogenesis (Porch and Jahn 2001), loss of pollen viability (Kafizadeh et al. 2008), poor pollen germination (Porch and Jahn 2001), and pollen tube growth inhibition (Kafizadeh et al. 2008). Further, the absence of pollen on stigma surface and loss of stigma receptivity (Jagadish et al. 2007), loss of ovule function (Gross and Kigel 1994), impaired fertilization (Dupuis and Dumas 1990), limited embryogenesis (Zinn et al. 2010), and reduced ovule number and increased ovule abortion lead to poor seed set (Young et al. 2004). The comparative sensitivity of reproductive stages, such as flowering and seed filling, to heat stress may vary according to the genotype (Sung et al. 2003).

Along with the reproductive damage, heat also causes numerous cellular abnormalities, such as alteration in the structure of proteins and enzymes (Demirevska-Kepova et al. 2005), inactivation of mitochondrial and chloroplast enzymes (Ashraf and Harris 2005), alteration in RNA and cytoskeleton structure, and finally cell death.

Oxidative stress is a common adverse effect of heat stress in cells because of the production of superoxides, lipid peroxides, and hydrogen peroxide (Yin et al. 2008). To counter the oxidative damage, the heat-stressed cells activate many enzymatic (superoxide dismutase, catalase, ascorbate peroxidase, glutathione reductase) and nonenzymatic (ascorbic acid, glutathione) antioxidants (Mittler 2002). Heat stress also affects nitrogen fixation and symbiosis in chickpea (Rodrigues et al. 2006). The high temperature (>32.5 °C) leads to a reduction in nodule formation and affects nodule structure and function (Roughley 1970; Kurdali 1996). Slightly increased day temperature (32.5 °C) delays nodulation and decreases nitrogen fixation in the plant and durability of the symbiotically active nodule population (Rawsthorne et al. 1985). At 35 °C, the nitrogenase activity gets affected in chickpea roots. The optimum soil temperatures for nodulation and nitrogen fixation for chickpea lie between 18° and 22 °C (Dart et al. 1975).

7.7.1 High Temperature and Alteration in Membrane Integrity

A link between a primary heat sensor and the phospholipid signaling events that follow a temperature stress remains to be determined. Heat shock activates phospholipid-based signaling pathways. Two key signaling lipids, phosphatidylinositol 4,5-bisphosphate (PIP2) and phosphatidic acid (PA), rapidly accumulate in plant cell membranes after the onset of a temperature stress. PIP2 is cleaved to generate inositol 1,4,5-trisphosphate (IP3) and diacylglycerol (DAG). IP3 also known as second messenger bind with IP3 receptors on the endoplasmic reticulum and release Ca^{2+} from that organelle into the cytoplasm. DAG activates protein kinase C (PKC). Under nonstressful conditions PIP2 is a minor component of cellular lipids, especially in plants where it represents 0.1% of total phospholipid pools. In plants, massive remodeling of the cytoskeleton is an early consequence of heat stress. PA activates a member of the PIPK family of enzymes, while PIP2 activate PLD.

The signaling cascade initiated by the heat-generated Ca^{2+} influx at the plasma membrane opens the channels and induces many critical components of the heat shock response (HSR). Activation of kinases and calmodulin and synthesis of HSPs may lead to induction of thermotolerance. Increase in PIPK and PLD activity is thought to be a secondary consequence of events initiated by the rise in cytoplasmic Ca^{2+} concentrations. Recent evidence suggests that plants efficiently phosphorylate IP3 to generate IP6 (phytic acid), which serve as second messenger for Ca^{2+} release and other functions. H2A. Z, a particular histone variant, mediates transcriptional changes in response to temperature (Horváth et al. 2012).

7.7.2 Heat Shock Proteins: The Molecular Chaperones

HSPs were first identified as the proteins induced by heat stress. HSR is a complex mechanism; hence, thermotolerance cannot be achieved by single heat shock transcription factor (HSF) or HSP gene. It is observed that nearly 2% genome is being affected by heat stress (Rizhsky et al. 2004). HSPs have been classified into five different families according to their molecular weight, viz., HSP100, HSP90, HSP70, HSP60, and sHSP. During heat stress HSP100 and sHSP play an important role. During heat stress, HSP70 and sHSP act as molecular chaperones (Larkindale et al. 2005). HSP100 is the member of AAA+ family of ATPase and participates in resolubilizing the protein aggregates. Cytosolic HSP100 is not essential for normal growth but is required for high-temperature tolerance (Bösl et al. 2006). Small HSP (sHSP) monomers are around 16–30 kDa, and most of the sHSPs form large oligomers in the native state. They have defined conserved carboxyl domain of about 90 amino acids, known as alpha crystalline domain. It is the most complex group of HSP in plants. It provides protection to almost all cellular compartments like nuclear-cytosolic compartment, chloroplasts, mitochondria, endoplasmic reticulum (ER), and peroxisomes (Nakamoto and Vigh 2007). In vitro, it was concluded that sHSPs bind to partially unfolded proteins in an ATP-independent manner and prevent the proteins from aggregation. Substrates that are denatured in the presence of sHSPs can be refolded and reactivated by Hsp70 and Hsp100 in few cases (Friedrich et al. 2004). In addition to chaperone function, sHSPs also modulate membrane fluidity and composition (Balogi et al. 2005). Among heat shock proteins, HSP101 play a specific role in acquired thermotolerance (Queitsch et al. 2000).

7.7.3 Heat Stress Transcription Factors

The transcription of HSP genes is controlled by regulatory proteins called HSFs, which serve as terminal component of signal transduction and play a prime role in heat stress response and in acquired thermotolerance (Kotak et al. 2007). HSF activity can be regulated at the transcriptional, posttranscriptional, and posttranslational levels. These posttranslational modifications can cause HSFs to oligomerize and translocate to the nucleus, bind to the promoters of HSR genes, and recruit histone acetyltransferase HAC1 (von Koskull-Döring et al. 2007). Plant HSFs include three conserved evolutionary classes, A, B, and C, which are mainly distinguished by the structural features of their oligomerization domains (Nover et al. 2001). Molecular mechanisms of plant HSF function were determined in tomato and *Arabidopsis*. In tomato, HsfA1a, HsfA2, and HsfB1 form a regulatory network and induce the expression of heat stress-responsive genes. HsfA1a is the master regulator and is constitutively expressed. It regulates the expression of HsfA2 and HsfB1. HsfA1a is the coactivator of HsfA2 which is a major HSF in thermotolerant cells. In tomato HsfB1 has been identified as a novel type of coactivator of class A HSFs and other transcription factors (Bharti et al. 2004). In *Arabidopsis* HsfA1a and HsfA1b are important for the initial phase of HS-responsive gene expression. HsfA2 controls expression under prolonged heat stress and recovery conditions (Busch et al. 2005). In contrast, in tomato the heat-induced expression of HsfA2 is not regulated by HsfA1a or HsfA1b. HsfA2 also regulate ascorbate peroxidase 2 (APX2), whereas HsfA4a and HsfA8 are hypothesized to act as sensors of ROS. It is regulated by DREB2A, a transcription factor involved in the regulation of dehydration-responsive genes in *Arabidopsis*. These findings suggest that HSF is involved in cross talk between heat stress and other abiotic stress signaling cascades (Sakuma et al. 2006). Fifty-two conserved HSFs have been reported in *Glycine max* regulated during heat stress (Scharf et al. 2012).

7.7.4 Signaling

Multiple signaling pathways are involved in HSR. Plants sense heat stress via different routes. The membrane is the first one to detect any change in the temperature in plant cell. Fluidity of the membrane affects the membrane protein and ROS accumulation, reduces energy levels, unfolds proteins or RNAs, and finally activates the heat sensor molecules (Saidi et al. 2011). Heat stress is also accompanied by some degree of oxidative stress. There occurs a cross talk between heat and oxidative stress signaling. HSFs sense H_2O_2 as a result of NADPH oxidase activity. These ROS induce HSP synthesis, but how ROS regulate heat stress-induced HSP expression is still unknown.

7.7.4.1 Calcium Signaling

Heat stress induces the influx of calcium inside the cell, which in turn activates the multiple signaling pathways in plants. Calcium influx serves as one of the primary heat sensors of plants (Reddy et al. 2011). CaM proteins are involved in the activation of transcription factors like WRKY and HSF. Calcium also activates several calcium-dependent protein kinases (CDPKs), which in turn activate multiple mitogen-activated protein kinases (MAPKs) or ROS-producing enzyme NADPH oxidase. When calcium binds to calmodulin, HSP90 mediate HSF phosphorylation. Phosphorylation of the key transcriptional regulator of thermotolerance MBF1c functions upstream to the dehydration-responsive element-binding (DREB) transcriptional activator and certain HSFs, which may be a direct or indirect result of CDPK activation (Mittler et al. 2012). Kinetic flux of cytosolic calcium is different under heat and cold shock. During cold shock, cytosolic ca2+ rise within minutes, whereas it is initiated in the recovery phase during heat shock (Plieth et al. 1999). Ca2+ transduces temperature-induced signals to MAPK (mitogen-activated protein kinase). HAMK was activated by heat shock, while during cold SAMK is activated (Sangwan et al. 2002).

7.7.4.2 Lipid Signaling

Heat stress results in the activation of phospholipase D (PLD) and phosphatidylinositol-4-phosphate 5-kinase (PIPK). It also governs the accumulation of various lipid signaling molecules such as phosphatidic acid, phosphatidylinositol 4,5-bisphosphate (PIP2), and D-myoinositol 1,4,5-trisphosphate (IP3). The accumulation of lipid signaling molecules can cause the opening of channels and the triggering of a calcium influx (Mishkind et al. 2009).

7.7.4.3 Unfolded Protein Response (UPR) in the ER and the Cytosol

Heat stress induces UPR pathway in endoplasmic reticulum (ER) and cytosol. The activation of the UPR pathway in plants involves the proteolytic cleavage and release of different bZIP transcription factors (TFs) from the ER membrane. The TFs enter the nuclei and activate the transcription of specific genes, which leads to the accumulation of ER chaperone transcripts and the activation of brassinosteroid signaling. In contrast to the ER, UPR, the cytosolic UPR, which is triggered by the presence of unfolded proteins in the cytosol, is primarily regulated by HSFA2, which binds to HSF-binding elements in the promoters of heat shock response genes (Che et al. 2010).

7.7.4.4 ROS Signaling

Heat stress-induced accumulation of ROS acts as signal to trigger the HSR. Heat stress survival and heat stress signal transduction require respiratory burst oxidase homolog D (RBOHD), a ROS-generating NADPH oxidase located in the plasma membrane. The activity of RBOHD is regulated by phosphorylation via calcium-dependent protein kinases (CDPKs) by direct binding of calcium to certain RBOHD. Therefore, it can be concluded that calcium activate RBOHD and result in the accumulation of ROS (Suzuki et al. 2011). The accumulated ROS might in turn activate downstream pathways via MBF1c, certain HSFs, MAPKs, and/or SnRKs, which might alter the redox state of the cell. ROS and calcium signaling are highly interlinked in the activation of the HSR (Mittler et al. 2004).

7.8 DNA Damage as an Effect of Heat Stress

DNA methylation regulates genes which are important for plant development and stress response. In plants, RNA-directed DNA methylation (RdDM) is directed by small RNAs. RdDM involves two plant-specific RNA polymerases PolIV and PolV and RNA-dependent RNA polymerase 2 (RDR2). Heat stress mobilizes retrotransposons making them transcriptionally active. They are synthesized in extra chromosomal DNA in *Arabidopsis* seedlings. Drought and salt stress induced CNG hypermethylation of the satellite DNA in the facultative halophyte *Mesembryanthemum crystallinum* that leads to metabolic switch from C3 to CAM photosynthesis mode (Naydenov et al. 2015). Epigenetic modifications such as DNA cytosine methylation, histone residue methylation, and acetylation also contribute to the adaptive response to the abiotic stress (Mizoi et al. 2012).

7.9 Drought and Heat Stress: Cross Talk

As sessile organisms, plants are constantly exposed to changes in temperature and other abiotic factors. Worldwide, extensive agricultural losses are attributed to heat often in combination with drought or other stresses (Mittler 2006). Important step in plant defense is the perception of stress on time. After recognizing the stress, plants activate basal defense mechanism, which further activate a series of signaling mechanism. Till now research has been mainly concentrated on understanding plant response to individual abiotic and biotic stresses. However, under field conditions, plants are often exposed to multiple stresses, and it has been found that exposure to one stress can strongly influence primary stress defense response of the plant to the other stress. The effect of the simultaneous stresses can be antagonistic, synergistic, or additive making the plants either more or less susceptible to the stress combination.

Exposure to abiotic/biotic stress also activates specific ion channels and kinase cascades like mitogen-activated protein kinase Ca^{2+}, ROS, phospholipid, mitochondrial functions, vesicle trafficking, and apoptosis (Ma and Bohnert 2007). Various phytohormones like abscisic acid (ABA), salicylic acid (SA), jasmonic acid (JA), and ethylene (ET) accumulate and reprogram the genetic machinery resulting in adequate defense reactions (Rejeb et al. 2014). TFs like WRKY, MYB, ERF, NAC, and HSFs are the master switches of cross talk among the various stresses. Epigenetic modifications such as DNA cytosine methylation, histone residue methylation, and acetylation also contribute to the adaptive response to the abiotic stress (Mirouze and Paszkowski 2011). Previously, it was also experimented in *Arabidopsis* that the gradual priming of plant with heat stress helps the plants to acquire thermotolerance, because gradual priming activates most of the transcript encoding molecular chaperones, ROS, as well as antioxidative enzymes (Mittler et al. 2012). It was also observed that cross-tolerance between environmental and biotic stress can induce positive effect and enhance resistance in plants which might lead to significant agricultural implication. Signaling molecules, common to both stresses, would enhance the resistance capabilities. Previous research has also been done on the use of chemicals for priming to promote cross-tolerance among the stresses (Rejeb et al. 2014).

7.10 ABA as Common Regulator

7.10.1 ABA Dependent

ABRE is the major cis element for ABA-responsive gene expression. AREB/ABFs are bZIP transcription factors which regulate ABA-dependent gene expression. Their transcriptional activities are controlled by the ABA-dependent phosphorylation. Perception of ABA and signal transduction includes SnRK2, group APP2C, and RCAR/PYR/PYL. Among them phosphorylation of SnRK2 is critical.

7.10.2 ABA Independent

Cis element sequence includes A/GCCGAC, which is designated as DRE/CRT. Two groups of AP2/ERF TFs were identified as DREB: DREB1/CBF and DREB2 in *Arabidopsis*. Overexpression of DREB2A also induces the expression of genes related to heat shock stress and improves thermotolerance in transgenic plants. Results indicate that DREB2A function in both dehydration and heat shock stress (Fig. 7.1).

A cross talk occurs between ABA-dependent and ABA-independent response pathways (Nakashima et al. 2014). Kumar et al. (2012) confirmed that ABA act upstream of the genes involved in osmolyte biosynthesis (proline, glycine betaine, and trehalose). Further confirmation was done by the exogenous application of ABA to chickpea seedling which produced osmolytes and caused much less oxidative damage in terms of MDA and H_2O_2 content.

In chickpea, APETALA2/ethylene response factor (AP2/ERF) superfamily and heat shock protein 90 (HSP90) are important classes of transcription factors and molecular chaperones which play an important role in abiotic and biotic stress. They also play an important role in various developmental processes. A total of 147 AP2/ERF and 5 HSP genes in chickpea have been found till date. Two DREB (Ca_16631 and Ca_02170) and three HSP90 genes (Ca_23016, Ca_09743, and Ca_25602) in chickpea were targeted as potential candidate gene for improving abiotic stress tolerance (Mizoi et al. 2012; Agarwal et al. 2016). Recent studies have also shown the relation between heat stress and DREB2A gene (Sato et al. 2014). Previously Rizhsky et al. (2004) showed that the combination of drought and heat stress at physiological level and molecular level is markedly different from the individual stress applied. The combination of drought and heat shock results in the

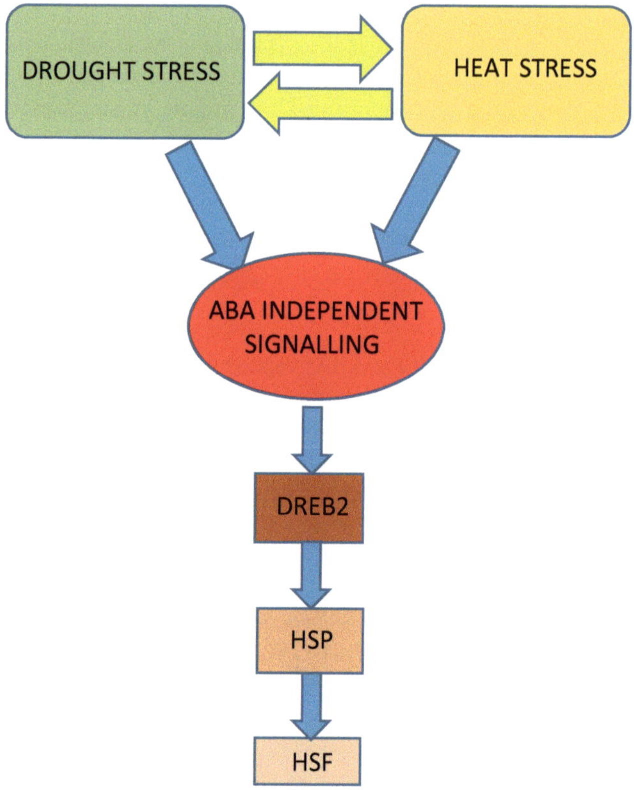

Fig. 7.1 Figure shows the ABA-independent pathway acting as a bridge between drought and heat stress pathway

activation of a unique genetic program that is different from that activated during drought or heat shock. Table 7.1 describes the common and individual response of plants.

7.11 Future Perspectives

Transcriptomics, proteomics, and metabolomics approaches have revealed plant responses under stress and their underlying mechanisms. Although complete genome sequence for chickpea is available, development of protein and metabolite databases will facilitate integration of data (Cabello et al. 2014). The use of omics in combination with forward genetic approaches would narrow down the candidate genes mainly responsible for the phenotypes and provide targets for functional characterization, further manipulation, and improvement of crops via genetic engineering. Another strategy includes engineering of regulatory network, which includes stress sensor proteins, ion channels, calcium-binding proteins, TF,

Table 7.1 Common and individual response of plants

Sl. No.	Plant processes	Drought stress response	Heat stress response	Common regulators for both the stress
1	Transcriptional factors	NAC, ABRE, CBF, MYC, ZF-HD	HSF, ERF	WRKY, MYB, ERF-HSF, DREB2
2	Signaling cascade	ABA dependent	Ethylene	ABA independent
			H_2O_2	
			Nitric oxide	ROS
			ABA dependent	Calcium
			ABA independent	
3	Protein level changes	LEA, dehydrins, RAB	HSF family (52 HSF in *Glycine max*)	Heat shock proteins (HSP)
		LTP, aquaporins, PRP		
4	Post translational changes	CDPK, AAPK, PKABA1, RPK1, PTPase	Phosphoproteins, phosphatases, HAMK	MAPK
5	Metabolomic changes	Proline, branched chain amino acid (BCAA), raffinose, oligosaccharides, gamma amino butyrate (GABA), and tricarboxylic acid (TCA), ABA	Antioxidative enzymes, NADPH, ATP, polyamine, GABA, quaternary amine, secondary metabolite	Proline, raffinose, galactinol, antioxidants, cytoskeleton changes

miRNA, and various hormones. Major limitation is transferring this approach from lab to the field. Exploring the potential of integrating synthetic biology approaches into current genetic engineering programs would open up new perspectives.

References

Abbo S, Berger J, Turner NC (2003) Viewpoint: evolution of cultivated chickpea: four bottlenecks limit diversity and constrain adaptation. Funct Plant Biol 30:1081–1087

Agarwal G, Garg V, Kudapa H, Doddamani D, Pazhamala LT, Khan AW, Thudi M, Lee SH, Varshney RK (2016) Genome-wide dissection of AP2/ERF and HSP90 gene families in five legumes and expression profiles in chickpea and pigeonpea. Plant Biotechnol J 14(7):1563

Alexandre C, Möller-Steinbach Y, Schönrock N, Gruissem W, Hennig L (2009) Arabidopsis MSI1 is required for negative regulation of the response to drought stress. Mol Plant 2:675–687

Araus J, Slafer G, Reynolds M, Royo C (2002) Plant breeding and drought in C3 cereals: what should we breed for? Ann Bot 89:925–940

Ashraf M, Harris PJ (2005) Abiotic stresses: plant resistance through breeding and molecular approaches. Food Prod Press, New York

Azimi SM, Kor NM, Ahmadi M, Shaaban M, Motlagh ZR, Shamsizadeh M (2015) Investigation of growth analysis in chickpea (*Cicer arietinum* L.) cultivars under drought stress. Int J Life Sci 9:91–94

Balogi Z, Török Z, Balogh G, Jósvay K, Shigapova N, Vierling E, Vígh L, Horváth I (2005) "Heat shock lipid" in cyanobacteria during heat/light-acclimation. Arch Biochem Biophys 436:346–354

Bartels D, Sunkar R (2005) Drought and salt tolerance in plants. Crit Rev Plant Sci 24:23–58

Basu P, Ali M, Chaturvedi S (2009) Terminal heat stress adversely affects chickpea productivity in northern India-Strategies to improve thermo tolerance in the crop under climate change. In: ISPRS Archives XXXVIII W3 workshop proceedings: impact of climate change on agriculture, 23:25

Bharti K, von Koskull-Döring P, Bharti S, Kumar P, Tintschl-Körbitzer A, Treuter E, Nover L (2004) Tomato heat stress transcription factor HsfB1 represents a novel type of general transcription coactivator with a histone-like motif interacting with the plant CREB binding protein ortholog HAC1. Plant Cell 16:1521–1535

Bösl B, Grimminger V, Walter S (2006) The molecular chaperone Hsp104—a molecular machine for protein disaggregation. J Struct Biol 156:139–148

Bray EA (1997) Plant responses to water deficit. Trends Plant Sci 2:48–54

Busch W, Wunderlich M, Schöffl F (2005) Identification of novel heat shock factor-dependent genes and biochemical pathways in *Arabidopsis thaliana*. The Plant J 41:1–14

Cabello JV, Lodeyro AF, Zurbriggen MD (2014) Novel perspectives for the engineering of abiotic stress tolerance in plants. Curr Opin Biotechnol 26:62–70

Campalans A, Messeguer R, Goday A, Pagès M (1999) Plant responses to drought, from ABA signal transduction events to the action of the induced proteins. Plant Physiol Biochem 37:327–340

Chanders S, Kumar PR, Bhadrarays S, Barmanl D (2008) Effect of increasing temperature on yield of some winter crops in northwest India. Curr Sci 94:82

Che P, Bussell JD, Zhou W, Estavillo GM, Pogson BJ, Smith SM (2010) Signaling from the endoplasmic reticulum activates brassinosteroid signaling and promotes acclimation to stress in Arabidopsis. Sci Signal 3:ra69–ra69

Chrispeels MJ, Maurel C (1994) Aquaporins: the molecular basis of facilitated water movement through living plant cells? Plant Physiol 105:9

Close TJ (1996) Dehydrins: emergence of a biochemical role of a family of plant dehydration proteins. Physiol Plant 97:795–803

Close T, Bray E (1993) Response of plants to cellular dehydration during environmental stress, The American Society of Plant Physiologists. Department of Botany and Plant Sciences, Riverside

Colmenero-Flores JM, Campos F, Garciarrubio A (1997) and Covarrubias*, A.A. Characterization of *Phaseolus vulgaris* cDNA clones responsive to water deficit: identification of a novel late embryogenesis abundant-like protein. Plant Mol Biol 35:393–405

Dart P, Islam R, Eaglesham A (1975) The root nodule symbiosis of chickpea and pigeonpea. In: Proceedings, international workshop in Grain Legumes, 13–16 January, ICRISAT, Patancheru

Demirevska-Kepova K, Holzer R, Simova-Stoilova L, Feller U (2005) Heat stress effects on ribulose-1, 5-bisphosphate carboxylase/oxygenase, Rubisco binding protein and Rubisco activase in wheat leaves. Biol Plant 49:521–525

Devasirvatham V, Tan D, Gaur P, Raju T, Trethowan R (2012) High temperature tolerance in chickpea and its implications for plant improvement. Crop Pasture Sci 63:419–428

Duke J (2012) Handbook of legumes of world economic importance. Springer, Dordrecht

Dupuis I, Dumas C (1990) Influence of temperature stress on in vitro fertilization and heat shock protein synthesis in maize (*Zea mays* L.) reproductive tissues. Plant Physiol 94:665–670

Ellis R, Lawn R, Summerfield R, Qi A, Roberts E, Chay P, Brouwer J, Rose J, Yeates S, Sandover S (1994) Towards the reliable prediction of time to flowering in six annual crops. V. Chickpea (*Cicer arietinum*). Exp Agric 30:271–282

Friedrich KL, Giese KC, Buan NR, Vierling E (2004) Interactions between small heat shock protein subunits and substrate in small heat shock protein-substrate complexes. J Biol Chem 279:1080–1089

Gaur P, Pande S, Sharma H, Gowda C, Sharma K, Crouch J, Vadez V (2007) Genetic enhancement of stress tolerance in chickpea: present status and future prospects. In: Crop production stress environments: genetic and management options. AGROBIOS International Publishing, Jodhpur, pp 85–94

Gross Y, Kigel J (1994) Differential sensitivity to high temperature of stages in the reproductive development of common bean (*Phaseolus vulgaris L.*). Field Crop Res 36:201–212

Holmberg N, Bülow L (1998) Improving stress tolerance in plants by gene transfer. Trends Plant Sci 3:61–66

Horváth I, Glatz A, Nakamoto H, Mishkind ML, Munnik T, Saidi Y, Goloubinoff P, Harwood JL, Vigh L (2012) Heat shock response in photosynthetic organisms: membrane and lipid connections. Prog Lipid Res 51:208–220

Hulse J (1989) Nature, composition and utilization of grain legumes. Uses of Tropical Grain Legumes 27: 11

Jagadish S, Craufurd P, Wheeler T (2007) High temperature stress and spikelet fertility in rice (*Oryza sativa* L.). J Exp Bot 58:1627–1635

Kader J-C (1997) Lipid-transfer proteins: a puzzling family of plant proteins. Trends Plant Sci 2:66–70

Kafizadeh N, Carapetian J, Kalantari KM (2008) Effects of heat stress on pollen viability and pollen tube growth in pepper. Res J Bio Sci 3:1159–1162

Kotak S, Larkindale J, Lee U, von Koskull-Döring P, Vierling E, Scharf K-D (2007) Complexity of the heat stress response in plants. Curr Opin Plant Biol 10:310–316

Kumar S, Kaushal N, Nayyar H, Gaur P (2012) Abscisic acid induces heat tolerance in chickpea (*Cicer arietinum* L.) seedlings by facilitated accumulation of osmoprotectants. Acta Physiol Plant 34:1651–1658

Kumar S, Thakur P, Kaushal N, Malik JA, Gaur P, Nayyar H (2013) Effect of varying high temperatures during reproductive growth on reproductive function, oxidative stress and seed yield in chickpea genotypes differing in heat sensitivity. Arch Agron Soil Sci 59:823–843

Kurdali F (1996) Nitrogen and phosphorus assimilation, mobilization and partitioning in rainfed chickpea (*Cicer arietinum* L.). Field Crop Res 47:81–92

Ladizinsky G, Adler A (1976) The origin of chickpea *Cicer arietinum* L. Euphytica 25:211–217

Larkindale J, Mishkind M, Vierling E (2005) Plant responses to high temperature. In: Plant abiotic stress, Blackwell, Oxford, pp 100–144

Leung J, Giraudat J (1998) Abscisic acid signal transduction. Annu Rev Plant Biol 49:199–222

Ma S, Bohnert HJ (2007) Integration of *Arabidopsis thaliana* stress-related transcript profiles, promoter structures, and cell-specific expression. Genome Biol 8:R49

Maestri E, Klueva N, Perrotta C, Gulli M, Nguyen HT, Marmiroli N (2002) Molecular genetics of heat tolerance and heat shock proteins in cereals. Plant Mol Biol 48:667–681

Meiri D, Breiman A (2009) Arabidopsis ROF1 (FKBP62) modulates thermotolerance by interacting with HSP90. 1 and affecting the accumulation of HsfA2-regulated sHSPs. The Plant J 59:387–399

Mirouze M, Paszkowski J (2011) Epigenetic contribution to stress adaptation in plants. Curr Opin Plant Biol 14:267–274

Mishkind M, Vermeer JE, Darwish E, Munnik T (2009) Heat stress activates phospholipase D and triggers PIP2 accumulation at the plasma membrane and nucleus. The Plant J 60:10–21

Mittler R (2002) Oxidative stress, antioxidants and stress tolerance. Trends Plant Sci 7:405–410

Mittler R (2006) Abiotic stress, the field environment and stress combination. Trends Plant Sci 11:15–19

Mittler R, Vanderauwera S, Gollery M, Van Breusegem F (2004) Reactive oxygen gene network of plants. Trends Plant Sci 9:490–498

Mittler R, Finka A, Goloubinoff P (2012) How do plants feel the heat? Trends Biochem Sci 37:118–125

Mizoi J, Shinozaki K, Yamaguchi-Shinozaki K (2012) AP2/ERF family transcription factors in plant abiotic stress responses. Biochimi Biophys Acta Gene Regul Mech 1819:86–96

Nakamoto H, Vigh L (2007) The small heat shock proteins and their clients. Cell Mol Life Sci 64:294–306

Nakashima K, Yamaguchi-Shinozaki K, Shinozaki K (2014) The transcriptional regulatory network in the drought response and its crosstalk in abiotic stress responses including drought, cold, and heat. Abiotic Stress: Mol Genet Genomics 5:25–31

Naydenov M, Baev V, Apostolova E, Gospodinova N, Sablok G, Gozmanova M, Yahubyan G (2015) High-temperature effect on genes engaged in DNA methylation and affected by DNA methylation in Arabidopsis. Plant Physiol Biochem 87:102–108

Nguyen KH, Van Ha C, Watanabe Y, Tran UT, Esfahani MN, Van Nguyen D, Tran L-SP (2015) Correlation between differential drought tolerability of two contrasting drought-responsive chickpea cultivars and differential expression of a subset of CaNAC genes under normal and dehydration conditions. Front Plant Sci 6

Nover N, Bharti K, Döring P, Mishra SK, Ganguli A, Scharf K-D (2001) Arabidopsis and the heat stress transcription factor world: how many heat stress transcription factors do we need? Cell Stress Chaperones 6:177–189

Peleg Z, Blumwald E (2011) Hormone balance and abiotic stress tolerance in crop plants. Curr Opin Plant Biol 14:290–295

Plieth C, Hansen UP, Knight H, Knight MR (1999) Temperature sensing by plants: the primary characteristics of signal perception and calcium response. The Plant J 18:491–497

Porch T, Jahn M (2001) Effects of high-temperature stress on microsporogenesis in heat-sensitive and heat-tolerant genotypes of *Phaseolus vulgaris*. Plant Cell Environ 24:723–731

Qin F, Sakuma Y, Tran L-SP, Maruyama K, Kidokoro S, Fujita Y, Fujita M, Umezawa T, Sawano Y, Miyazono K-i (2008) Arabidopsis DREB2A-interacting proteins function as RING E3 ligases and negatively regulate plant drought stress–responsive gene expression. Plant Cell 20:1693–1707

Qu A-L, Ding Y-F, Jiang Q, Zhu C (2013) Molecular mechanisms of the plant heat stress response. Biochem Biophys Res Commun 432:203–207

Queitsch C, Hong S-W, Vierling E, Lindquist S (2000) Heat shock protein 101 plays a crucial role in thermotolerance in Arabidopsis. Plant Cell 12:479–492

Rampino P, Pataleo S, Gerardi C, Mita G, Perrotta C (2006) Drought stress response in wheat: physiological and molecular analysis of resistant and sensitive genotypes. Plant Cell Environ 29:2143–2152

Rawsthorne S, Hadley P, Summerfield R, Roberts E (1985) Effects of supplemental nitrate and thermal regime on the nitrogen nutrition of chickpea (*Cicer arietinum* L.). Plant Soil 83:279–293

Reddy AS, Ali GS, Celesnik H, Day IS (2011) Coping with stresses: roles of calcium-and calcium/calmodulin-regulated gene expression. Plant Cell 23:2010–2032

Rejeb IB, Pastor V, Mauch-Mani B (2014) Plant responses to simultaneous biotic and abiotic stress: molecular mechanisms. Plants 3:458–475

Rizhsky L, Liang H, Shuman J, Shulaev V, Davletova S, Mittler R (2004) When defense pathways collide. The response of Arabidopsis to a combination of drought and heat stress. Plant Physiol 134:1683–1696

Rodrigues CS, Laranjo M, Oliveira S (2006) Effect of heat and pH stress in the growth of chickpea mesorhizobia. Curr Microbiol 53:1–7

Roughley R (1970) The influence of root temperature, Rhizobium strain and host selection on the structure and nitrogen-fixing efficiency of the root nodules of *Trifolium subterraneum*. Ann Bot 34:631–646

Saidi Y, Finka A, Goloubinoff P (2011) Heat perception and signalling in plants: a tortuous path to thermotolerance. New Phytol 190:556–565

Sakuma Y, Maruyama K, Qin F, Osakabe Y, Shinozaki K, Yamaguchi-Shinozaki K (2006) Dual function of an Arabidopsis transcription factor DREB2A in water-stress-responsive and heat-stress-responsive gene expression. Proc Natl Acad Sci 103:18822–18827

Sangwan V, Örvar BL, Beyerly J, Hirt H, Dhindsa RS (2002) Opposite changes in membrane fluidity mimic cold and heat stress activation of distinct plant MAP kinase pathways. The Plant J 31:629–638

Santner A, Estelle M (2009) Recent advances and emerging trends in plant hormone signalling. Nature 459:1071–1078

Sato H, Mizoi J, Tanaka H, Maruyama K, Qin F, Osakabe Y, Morimoto K, Ohori T, Kusakabe K, Nagata M (2014) Arabidopsis DPB3-1, a DREB2A interactor, specifically enhances heat stress-induced gene expression by forming a heat stress-specific transcriptional complex with NF-Y subunits. Plant Cell 26:4954–4973

Saxena N, Saxena M, Johansen C, Virmani S, Harris H (1996) Adaptation of chickpea in the West Asia and North Africa region. ICRISAT, Aleppo

Scharf KD, Berberich T, Ebersberger I, Nover L (2012) The plant heat stress transcription factor (Hsf) family: structure, function and evolution. Biochim Biophys Acta 1819:104–119

Seki M, Umezawa T, Urano K, Shinozaki K (2007) Regulatory metabolic networks in drought stress responses. Curr Opin Plant Biol 10:296–302

Singh K, Malhotra R, Halila M, Knights E, Verma M (1994) Current status and future strategy in breeding chickpea for resistance to biotic and abiotic stresses. In: Expanding the production and use of cool season food legumes. Springer, pp 572–591

Summerfield R, Hadley P, Roberts E, Minchin F, Rawsthorne S (1984) Sensitivity of Chickpeas (*Cicer arietinum*) to Hot Temperatures during the Reporductive Period. Exp Agric 20:77–93

Sung D-Y, Kaplan F, Lee K-J, Guy CL (2003) Acquired tolerance to temperature extremes. Trends Plant Sci 8:179–187

Suzuki N, Miller G, Morales J, Shulaev V, Torres MA, Mittler R (2011) Respiratory burst oxidases: the engines of ROS signaling. Curr Opin Plant Biol 14:691–699

Urano K, Kurihara Y, Seki M, Shinozaki K (2010) 'Omics' analyses of regulatory networks in plant abiotic stress responses. Curr Opin Plant Biol 13:132–138

Van der Maesen L (1972) A monograph of the genus with special reference to the chickpea (*C. arietinum* L.) Veeman and Zoren. Wageningen

Varshney RK, Song C, Saxena RK, Azam S, Yu S, Sharpe AG, Cannon S, Baek J, Rosen BD, Tar'an B (2013) Draft genome sequence of chickpea (*Cicer arietinum*) provides a resource for trait improvement. Nat Biotechnol 31:240–246

von Koskull-Döring P, Scharf K-D, Nover L (2007) The diversity of plant heat stress transcription factors. Trends Plant Sci 12:452–457

Wery J, Silim S, Knights E, Malhotra R, Cousin R (1993) Screening techniques and sources of tolerance to extremes of moisture and air temperature in cool season food legumes. Euphytica 73:73–83

Wu Y, Kuzma J, Maréchal E, Graeff R, Lee HC, Foster R, Chua N-H (1997) Abscisic acid signaling through cyclic ADP-ribose in plants. Science 278:2126–2130

Yin H, Chen Q, Yi M (2008) Effects of short-term heat stress on oxidative damage and responses of antioxidant system in *Lilium longiflorum*. Plant Growth Regul 54:45–54

Young LW, Wilen RW, Bonham-Smith PC (2004) High temperature stress of *Brassica napus* during flowering reduces micro-and megagametophyte fertility, induces fruit abortion, and disrupts seed production. J Exp Bot 55:485–495

Zhu B, Choi D-W, Fenton R, Close T (2000) Expression of the barley dehydrin multigene family and the development of freezing tolerance. Mol Gen Genet 264:145–153

Zinn KE, Tunc-Ozdemir M, Harper JF (2010) Temperature stress and plant sexual reproduction: uncovering the weakest links. J Exp Bot:erq053

Interaction of Light and Temperature Signaling at the Plant Interphase: From Cue to Stress

8

Juhi Bhattacharya, Upendra Kumar Singh, and Aashish Ranjan

Abstract

Light and temperature are two of the most important environmental stimuli regulating plant growth and development. While majority of the studies so far investigated light and temperature signaling in isolation, an understanding of the interaction of the two pathways is essential to generate a holistic view of plant responses to environmental cues. Recent studies have demonstrated significant overlap between the two signaling pathways at the molecular level to optimize plant growth as evident by involvement of phytochromes and downstream transcription factors in both the pathways. Plants respond to limited light availability and high ambient temperature via similar phenotypic response, such as stem elongation. Moreover, the existence of shared signaling components, such as common plant hormones and transcription factors, for the two pathways has also been deciphered. Together, the studies showing a significant overlap between light and temperature signaling cascades at the molecular level that involves an array of transcription factors and phytohormone-related genes suggest a complex cross talk between the two signaling pathways, which is fundamental to plant growth in natural environments.

Keywords

Light • Temperature • Signaling • Stress • Plant development

J. Bhattacharya • U.K. Singh • A. Ranjan (✉)
National Institute of Plant Genome Research, Aruna Asaf Ali Marg, 10531, New Delhi 110 067, India
e-mail: aranjan@nipgr.ac.in

M. Senthil-Kumar (eds.), *Plant Tolerance to Individual and Concurrent Stresses*, DOI 10.1007/978-81-322-3706-8_8

8.1 Introduction

In their natural habitat, plants are exposed to multiple environmental factors, such as light, temperature, humidity, precipitation, and nutrient availability, throughout their life cycle. As sessile photoautotrophs, plants need to adjust their growth, development, and physiology with the changing environmental conditions. Among the various environmental factors, light and temperature are arguably the most important and well-studied factors influencing plant life cycle (Franklin et al. 2014; Heggie and Halliday 2005; Jin et al. 2011; Whitelam and Halliday 2007). Light and temperature not only serve as cues to initiate multiple developmental processes, but are also required for optimal performances of biochemical and physiological cascades for the successful plant life cycle (Penfield 2008; Kami et al. 2010). An optimal range of light and temperature, which varies for different plant species, is critical for the proper growth and development. Sub- and supraoptimal conditions of these environmental factors are unfavorable to growth and development of plants. For example, plants respond to the limited light availability through a set of developmental responses, such as stem elongation, reduced leaf area, early flowering, and reduced biomass, that contribute to shade avoidance (Franklin 2008; Casal 2013). Similarly plants also respond to higher ambient temperature through a suit of developmental changes, similar to shade avoidance (Franklin et al. 2011; Gray et al. 1998; Koini et al. 2009; Proveniers and van Zanten 2013). Plants have evolved multiple receptors to perceive diverse environmental cues and changes in environmental conditions, and they transduce the information to downstream metabolic or biochemical changes in order to enable plants to survive under given conditions. Understanding the genetic and molecular basis of perception of environmental signals is one of the first steps in deciphering the environmental signaling cascade in plants. Studies on the downstream responses of environmental signals have strongly suggested that most, if not all, environmental stimuli act at least in part through modifying gene expression through transcription factors and hormonal activities (Stavang et al. 2009; Jiao et al. 2007; Alabadi and Blazquez 2009). Moreover different environmental signaling pathways may interact to shape final growth and development of plants. For example, the developmental responses of plants to shade (limited light) and high-temperature conditions show a large overlap at both phenotypic and genetic levels (Franklin et al. 2011, 2014; Koini et al. 2009). Such cross talk of signaling pathways might have evolutionary advantage as plants may experience multiple stresses and the interaction of the pathways with common master regulators shall equip plants to survive under multiple stress situations.

In this chapter we provide a brief overview of light and temperature perception and signaling in plants. Further, we discuss the cross talk of light and temperature signaling to regulate plant developmental responses under unfavorable conditions of shade and high temperature.

8.2 Light as an Environmental Cue for Plant Growth and Development

8.2.1 Light Perception and Signaling

Plants perceive light quality, quantity, direction, and periodicity to modulate a plethora of developmental responses, such as seed germination and seedling establishment to mature plant architecture and the onset of reproductive development. In addition light conditions also influence metabolism, such as the biosynthesis of chlorophyll and anthocyanins (Kami et al. 2010; Whitelam and Halliday 2007). The effect of light is evident early in the plant life cycle as seedlings grown in darkness exhibit skotomorphogenesis, a characteristic set of morphological features that includes long hypocotyls, prominent apical hook, and closed cotyledon without chlorophyll. In contrast, light signal induces photomorphogenesis resulting in reduced hypocotyl length and open green cotyledons (Kendrick and Kronenberg 1994; Leivar and Quail 2011; Han et al. 2007; Wu 2014). A variety of photoreceptors, such as phytochromes, cryptochromes, phototropins, Zeitlupe family of receptors, and UVR8, have been identified in plants for perceiving ambient light conditions and to initiate light signal transduction (Moglich et al. 2010; Franklin and Whitelam 2004; Franklin et al. 2005). Each of these photoreceptors perceives specific light conditions. The phytochrome family of photoreceptors perceives red (R) and far-red (FR) wavelengths. Blue (B) wavelength is perceived by cryptochromes, phototropins, and Zeitlupe (ZTL) protein family (ZTL/LKP2/FKF1). Cryptochromes and phototropins also detect ultraviolet (UV)-A wavelengths, whereas UV-B is perceived by UVR8 (Franklin and Whitelam 2004; Lin and Shalitin 2003; Li and Yang 2007; Christie 2007; Franklin and Quail 2010; Kim et al. 2007; Tilbrook et al. 2013; Somers et al. 2004; Rizzini et al. 2011).

There are two basic molecular mechanisms by which photoreceptors regulate gene expression and downstream response to induce photomorphogenesis. First mechanism involves alteration of gene expression rapidly in response to light through modulation of the activity of transcription factors by activated photoreceptors. This may involve direct interaction of photoreceptors with the transcription factors. One such example is phytochrome-mediated degradation of PHYTOCHROME-INTERACTING FACTOR (PIF) family of transcription factors (Leivar and Monte 2014; Leivar and Quail 2011; Duek and Fankhauser 2005). PIFs are bHLH transcription factors that act as repressors of light signaling. Seven members of PIF family of transcription factors [PIF1/PIF3-LIKE 5 (PIL5), PIF3, PIF4, PIF5/PIL6, PIF6/PIL2, PIF7, and PIF8] have been shown to interact physically with phytochromes. Phytochromes are synthesized in their inactive R-absorbing (Pr) form, and biological activity is acquired upon photo-conversion to the FR-absorbing (Pfr) form (Kendrick and Kronenberg 1994). Light-mediated activation of phytochromes also triggers translocation of the phytochromes from the cytoplasm to the nucleus, where active phytochromes physically interact with

PIFs to promote their ubiquitination and degradation (Al-Sady et al. 2006; Bauer et al. 2004; Castillon et al. 2007; Kircher et al. 2002; Martinez-Garcia et al. 2000; Nagatani 2004; Shen et al. 2008). Recent studies also show physical interaction of cryptochromes with PIF transcription factors to regulate light-dependent processes (Pedmale et al. 2016). Moreover cryptochromes also directly interacts with other transcription factors, such as bHLH transcription factor CIB1, to regulate flowering time (Liu et al. 2008).

Regulated protein degradation is second major mechanism underlying light signal transduction pathway (Hoecker 2005; Henriques et al. 2009). Downstream to photoreceptors, a group of CONSTITUTIVELY PHOTOMORPHOGENIC/DE-ETIOLATED/FUSCA (COP/DET/FUS) proteins are negative regulators of light control of plant development. These proteins repress photomorphogenesis by targeting transcription factors required for light signaling, such as HFR1, HY5, HYH, and LAF1, for degradation in darkness (Chory et al. 1989; Duek et al. 2004; Kim et al. 2002; Osterlund et al. 2000; Yang et al. 2005). COP1, the most well-characterized protein among the *COP/DET/FUS* genes, forms active E3 ubiquitin ligases in combination with SUPPRESSOR OF PHYTOCHROME-A (SPA) proteins (Zhu et al. 2008; Yi and Deng 2005; Hoecker and Quail 2001; Deng et al. 1992). COP1-SPA complex promotes the ubiquitination followed by subsequent degradation of the activators of the light response (Osterlund et al. 2000; Seo et al. 2003; Yang et al. 2005). In the light, activated photoreceptors inhibit the functions of COP/DET/FUS proteins including COP1-SPA complex through the physical interaction with the complex, and hence the transcription factors required for light signaling are stabilized to induce normal photomorphogenesis (Hoecker 2005; Henriques et al. 2009). Seedlings with mutations in any of these genes show constitutive photomorphogenesis in darkness in displaying short hypocotyls and open cotyledons. These mutants have strongly elevated levels of key photomorphogenesis-promoting transcription factors such as HY5 and HFR1 in darkness (Laubinger et al. 2004; Yi and Deng 2005). COP1 and SPA genes also regulate other developmental stages, such as adult plant development and photoperiodic flowering. The regulation of photoperiodic flowering by the COP1-SPA complex takes place through the regulation of stability of transcription factor CONSTANS, whereas the substrates for adult plant development through the complex have not been identified yet (Ishikawa et al. 2006; Jang et al. 2008; Laubinger et al. 2006).

8.2.2 Light Regulation of Plant Development

The photoreceptors show redundant, synergistic, and sometimes mutually antagonistic mechanisms of action to regulate plant developmental and physiological processes (Sullivan and Deng 2003; Franklin et al. 2005; Wang and Wang 2015). Phytochromes are the sole photoreceptors inducing seed germination in *Arabidopsis*, whereas a complex interplay of both phytochromes and cryptochromes is implicated in seedling de-etiolation by sensing specific light quality and quantity (Reed et al. 1993; Nagatani et al. 1993; Hennig et al. 2002; Casal and Sanchez 1998). Physical interactions between the photoreceptors, such as cry1 and phyA and cry2 and phyB, have been demonstrated to influence the seedling development in response to light (Ahmad et al. 1998; Mas et al. 2000). Moreover the combined action of phyA, phyB, cry1, and cry2 regulates the stimulation of chlorophyll synthesis by light (McCormac and Terry 2002). The effect of photoreceptors in the regulation of adult plant and leaf development is best characterized in the shade-avoidance syndrome, which results due to shading by neighboring vegetation. Plants respond to this limited light condition that is enriched in far-red wavelength, by stimulated elongation growth, reduced leaf development, increased apical dominance, and reduced branching (Franklin 2008). phyB plays a predominant role in this process along with the contribution from phyD and phyE (Franklin et al. 2003). Further, phyA, phyB, and phyE function redundantly to maintain the compact rosette habit of *Arabidopsis* (Devlin et al. 1998). Photoreceptors, phyA, phyB, and cry2, regulate photoperiodic flowering by regulating the stability of transcription factor CONSTANS (CO) that activates transcription of *FLOWERING LOCUS T (FT)* (Imaizumi and Kay 2006; Mockler et al. 2003; Valverde et al. 2004). Thus, the multiplicity of responses to environmental light signals available to plants results from the combined action of all photoreceptors that permit an array of developmental responses under changing light conditions.

8.3 Temperature as an Environmental Cue for Plant Growth and Development

8.3.1 Ambient Temperature Perception and Signaling

Temperature is another important seasonal cue that not only acts as a stimulus to control the timing of developmental transitions but also enables plants to predict and consequently prevent the adverse effect of future temperature extremes. Similar to light stimulus, temperature also regulates almost all stages of plant growth and development (Wigge 2013; Penfield 2008). However, molecular mechanisms underlying temperature perception and signaling in plants have been harder to resolve compared to light signaling. This is largely due to complex and ubiquitous effects of temperature on cellular responses. There have been insights into the temperature-sensing mechanisms, but the bona fide temperature receptors have not been identified so far. Ca^{+2} ion appears to be an important component of temperature perception and signaling in plants. CNGC2 (a calcium-conducting cyclic nucleotide-gated channel) has been shown to have a thermosensory function to induce the expression of *heat shock protein* (*HSP*) in response to high temperatures (Finka et al. 2012). High-temperature-induced membrane fluidity leads to opening of these membrane Ca^{+2} channels, facilitating the high-temperature response. High-temperature-mediated membrane fluidity not only activates Ca^{+2} signaling but also modulates lipid signaling. Phospholipase D (PLD) and phosphatidylinositol-4-phosphate 5-kinase, components of lipid signaling, are activated by heat stress contributing to changes in membrane composition in response to high temperature (Mishkind et al. 2009; Xiao et al. 2011; Yao et al. 2011). In addition, cold treatment has also been shown to activate calcium-permeable channels in *Arabidopsis* mesophyll cells (Carpaneto et al. 2007).

Most of the studies till date have concentrated on temperature extremes, with limited information toward the elucidation of the signaling mechanisms underlying plant responses to small changes in ambient temperature. Chromatin remodeling has, recently, appeared to be a critical regulator of gene expression pattern in response to changes in ambient temperature (Wigge 2013). A recent study identified the role of H2A.Z nucleosomes at specific promoters in coordinating the changes in gene expression pattern in response to ambient temperature changes (Wigge 2013; Kumar and Wigge 2010). H2A.Z nucleosome occupancy, which has been shown to prevent gene expression, is proposed to be rate limiting for the expression of temperature-responsive genes. With increasing temperature, H2A.Z nucleosomes are selectively evicted from the promoters of temperature-responsive genes for opening up access for transcription factors to activate the expression of those genes (Kumar and Wigge 2010). Consistent with this, mutation in *actin-related protein 6 (ARP6)* gene, which is responsible for installing H2A.Z into nucleosomes in place of H2A, exhibits failure in inserting H2A.Z in the nucleosome resulting in constitutive expression of warm-temperature-responsive genes (Kumar and Wigge 2010). Dislodging H2A.Z from nucleosomes appears to induce the expression *HSP* (*heat shock protein*) genes in response to higher temperatures (Mittler et al. 2012). PIF4 is a critical regulator for high ambient temperature signaling (Franklin et al. 2014), and reduced H2A.Z occupancy has been shown to induce the accessibility of PIF4 to the binding sites of target genes to show high-temperature response (Kumar et al. 2012).

8.3.2 Temperature Regulation of Plant Development

The need of an optimal temperature range for plant is obvious at the very beginning of its life cycle as many plants require a period of cold to promote its germination. The pronounced effect of temperature on overall plant growth has also been demonstrated. For example, *Arabidopsis* plants grown at 22 °C displayed the greatest leaf area and biomass. Plants grown at 16 °C resulted in dwarfed plants with compact rosette, whereas plants at 28 °C showed significant petiole elongation, leaf hyponasty, and reduced leaf area, comparable to the shade-avoidance syndrome (Atkin et al. 2006; Franklin 2008; Gray et al. 1998). Increase in temperature has been shown to shorten the time required to produce a leaf, suggesting temperature promotes the leaf initiation and development (Lopez and Runkle 2004). Moreover temperature is a potent regulator of flowering time. Many plant species require a prolonged exposure to cold, a process called vernalization, to accelerate their flowering (Henderson and Dean 2004). Following vernalization, exposure to high temperatures (27 °C) promotes flowering via a thermosensory role proposed for the floral repressor FLOWERING LOCUS M (FLM) (Balasubramanian et al. 2006; Blazquez et al. 2003). Moreover, small changes in ambient temperature can regulate flowering time through the modulation of FLOWERING LOCUS T (FT) in a thermosensory pathway (Lee et al. 2007).

8.4 Unfavorable Light and Temperature Conditions: Stress for Plants

Unfavorable light conditions, including intensity, quality, as well as periodicity, adversely affect plant metabolism and growth. Both high and low light intensities limit the overall plant performance. Excessive light intensity leads to photodamage of the leaf and reduces the crop yield (Ort 2001; Galvez-Valdivieso et al. 2009). High light intensity, in excess of photosynthetic capacity, dramatically increases the photosynthetic generation of biologically damaging molecules including reduced and excited species of oxygen, peroxides, radicals, and triplet state excited pigments that cause oxidative damage to the photosynthetic apparatus, photobleaching, and cell death (Karpinski et al. 1999; Mittler et al. 2004; Triantaphylides et al. 2008; Asada 1996). Plants tend to escape photodamage and activate photoprotection to dissipate excess light, detoxify photosynthetically produced reactive molecules, and induce a variety of repair processes (Niyogi 1999). Non-photochemical quenching is one such mechanism toward the thermal dissipation of excess light intensity (Ort 2001; Muller et al. 2001; Eberhard et al. 2008). Additional dissipation of excitation energy is also achieved by photochemical quenching (Baker et al. 2007). Furthermore, plants also get acclimatized to high light intensity and prevent the damaging effects of ROS through the activities of various antioxidant enzymes, such as ascorbate peroxidase, catalase, and superoxide dismutase, and phytohormone ABA (Ball et al. 2004; Fryer et al. 2003; Galvez-Valdivieso et al. 2009). Excess light intensity may also influence stomatal conductance and water transport capacity, likely due to combined effect of high light intensity and associated increase in temperature (Farquhar and Sharkey 1982; Araujo et al. 2011). The combined effects of photoreceptors, phototropins, cryptochromes, and phytochrome B, on stomatal density and/or opening, increase stomatal conductance in response to high light intensity (Boccalandro et al. 2012; Kinoshita et al. 2001; Mao et al. 2005). Similarly phyA and phyB enhance xylem development favoring water transport to leaves under high light intensity (Casal et al. 1994; Auge et al. 2012).

High temperature is one of the major abiotic stresses for plants that affect photosynthesis, yield, and productivity (Hasanuzzaman et al. 2013; Bita and Gerats 2013). High temperature affects both metabolic and physiological processes that result in wide array of responses, such as programmed cell death of leaves, delayed flowering, and sometimes death of the whole plant (McClung and Davis 2010; Mittler et al. 2012; Suzuki et al. 2012). Transfer of plants to extremely high temperatures initiates the expression of heat shock protein (HSP)/chaperone cascades which prevent the misfolding, denaturation, and aberrant aggregation of cellular proteins (Queitsch et al. 2000; Su and Li 2008; Yamada et al. 2007; Dafny-Yelin et al. 2008). Stressfully high temperature changes the membrane fluidity that results in Ca^{+2} influx triggering respiratory burst oxidase homolog D (RBOHD), which leads to ROS accumulation and programmed cell death in cells (Miller et al. 2009; Mittler et al. 2004; Saidi et al. 2010; Suzuki et al. 2012). In addition, other regulatory proteins and antioxidant enzymes are involved in response to heat stress (Mittler et al. 2012; Hasanuzzaman et al. 2013). Though the effect of heat stress on the plants has been observed to be drastic, slight changes in ambient temperature also significantly alter the plant growth and development (McClung and Davis 2010; Wigge 2013). Here we will mostly focus on the plant developmental responses to changes in the ambient temperature, temperature range below the heat-stress, and underlying genetic basis.

8.5 Interaction of Light and Temperature Signaling: Shade and High Ambient Temperature

Considering light and temperature as two most important environmental signals and that they regulate almost all stages of plant growth and development, the interaction between the two signaling pathways is inevitable. Recent studies have demonstrated a complex cross talk between light and temperature signals in the regulation of germination, plant architecture, flowering, and the enhancement of freezing tolerance (Franklin et al. 2014). The functions of different phytochromes strongly depend on the ambient temperature, suggesting the strong cross talk between the light and temperature signaling (Franklin and Quail 2010). Moreover, some photoreceptor mutants fail to show adult plant and leaf phenotype when grown under cooler temperature (Halliday et al. 2003). The integration of phytochrome and temperature signaling pathways has been reported in the regulation of multiple developmental processes, including germination (Donohue et al. 2007; Heschel et al. 2007; Penfield 2008) and flowering (Halliday et al. 2003; Halliday and Whitelam 2003). The individual contributions of the photoreceptors for germination of *Arabidopsis* seeds are determined by the temperature conditions. PhyB mediates germination under broad range of temperature, whereas PhyA and PhyE induce germination at cooler and warmer temperatures, respectively (Heschel et al. 2007). Similarly, transition from vegetative to reproductive development depends upon the compounding effects of light (day-length) and low temperature (Reeves and Coupland 2000). Cold acclimation in plants is also dependent upon photoreceptors. Transcriptomic analysis of plants treated with low red/far-red wavelengths at 16 ° and 22 °C revealed ambient temperature-dependent, light-quality regulation of the C-repeat-binding factor (CBF) regulon, a suite of genes involved in cold acclimation, and the acquisition of freezing tolerance (Franklin and Whitelam

2007). Photoreceptors, PhyB and PhyD, regulate the expression of *CBF* and its downstream target *COLD-REGULATED (COR)*, genes that are critical for cold acclimation, at low temperatures (Franklin and Whitelam 2007). Limited light availability, i.e., shade, and high ambient temperature are shown to induce similar phenotypic effects in the model plant *Arabidopsis* via the involvement of common transcription factors and phytohormone responses (Franklin et al. 2014, 2011; Koini et al. 2009). In the next sections, we will primarily focus on cross talk of light and temperature signaling for the control of adult plant development with special reference to shade and high ambient temperature.

8.5.1 High Similarity in Plant Response to Shade and High Ambient Temperature

One of the best examples for influence of light conditions on adult plant development is shade avoidance, a condition in which plants respond to limited light conditions through developmental alterations. The light quality required for optimal plant growth and development is enriched with red wavelength. However, under the canopy of nearby vegetation, plants receive light enriched with far-red wavelength and thus low red/far-red ratio and reduced photosynthetically active radiation (PAR) (Casal 2012). This is because the surrounding canopy absorbs most of red and blue light. The plants respond to this suboptimal light conditions by exhibiting shade-avoidance syndrome (SAS) which is characterized by elongation of the stem and petiole, erect leaf, premature flowering, and increased apical dominance (Casal 2013; Franklin 2008).

The set of morphological and architectural changes induced by higher ambient temperatures, below the heat-stress range, is collectively referred as thermomorphogenesis. The signature feature of thermomorphogenesis is elongation of the hypocotyl, stem, and petiole (Raschke et al. 2015; Zhu et al. 2015; Miyazaki et al. 2015; Lee et al. 2014; Box et al. 2015; Gray et al. 1998). The reasoning behind high-temperature-induced elongation response has been proposed as a way to move the sensitive meristematic and photosynthetically active tissues away from heat-absorbing soil and to promote cooling by allowing better access to moving air (Gray et al. 1998). In addition, plants also show strong hyponastic growth in response to high temperature. The plants adapted to high ambient temperature also have fewer stomata and develop smaller and thinner leaves. These phenotypic features help plants to mitigate high-temperature effects by enhancing evaporative cooling through increased surface area (Crawford et al. 2012; Vasseur et al. 2011). Moreover, increase in ambient temperature also promotes early flowering in *Arabidopsis* (Balasubramanian et al. 2006).

Thus, the plants respond to both high ambient temperature and shade conditions through a common set of phenotypic responses that include elongated stem, hypocotyl, and petioles in addition to reduced leaf area and early flowering.

8.5.2 Strong Parallels in Molecular Responses to Shade and High Ambient Temperature

8.5.2.1 Molecular Mechanisms Underlying Shade-Avoidance Response

The plants respond to the enriched far-red environment of shade through phytochrome-PIF module. PhyB is the major photoreceptor for attenuating the shade-avoidance syndrome under high red/far-red of normal light as the *phyB* mutant displays constitutive shade avoidance in the form of exaggerated elongation response and early flowering (Nagatani et al. 1991; Reed et al. 1993). phyA, phyD, and phyE also regulate SAS redundantly, albeit with minor contributions (Franklin et al. 2003). Under shade conditions, phytochromes predominate in inactive Pr form due to high far-red environment leading to stabilization/activation of phytochrome-interacting factors (PIFs), in particular PIF3, PIF4, PIF5, and PIF7, which collectively promotes elongation growth. In addition homeobox transcription factor ATHB2/HAT4 has also been associated as a positive regulator of growth during SAS (Carabelli et al. 1996; Salter et al. 2003). This elongation growth helps plants to come out from the shade of neighbors and have better access to optimal light conditions. The shade avoidance can be detrimental for crop yield as carbon resources are directed for the elongation growth at the expense of biomass production (Casal 2012).

PIF family of bHLH transcription factors acts as the primary hub for a signaling cascade to promote cell elongation (de Lucas and Prat 2014; Carriedo et al. 2016). Downstream to phytochrome-PIF module, phytohormones, particularly auxin, are important determinant of increased elongation growth in response to shade. PIF transcription factors, stabilized/activated in response to shade conditions, directly bind the promoter of auxin biosynthetic genes, such as *TRYPTOPHAN AMINOTRANSFERASE OF ARABIDOPSIS 1 (TAA1)* and *YUCCA8 (YUC8)*, and auxin-responsive *AUX/IAA* genes to increase their expression, resulting in a large increase in free auxin in the cotyledons and leaves (Fig. 8.1) (Lorrain et al. 2008; Hornitschek et al. 2012). It is believed that auxin produced in the cotyledons/leaves are transported into the hypocotyls/petiole to promote hypocotyl and petiole elongation in response to shade (Hornitschek et al. 2012; Lorrain et al. 2008; Leivar and Quail 2011; Morelli and Ruberti 2002; Procko et al. 2014). Auxin modulates cell-wall remodeling and cell elongation via regulation of expansins and xyloglucan endotransglucosylase/hydrolases (XTHs) (Sasidharan et al. 2010). In addition, modulation of biosynthesis and activities of other plant hormones, such as gibberellins, brassinosteroids, cytokinins, and ethylene, are also important for shade-avoidance response. Direct PIF interaction with DELLA proteins links gibberellin and brassinosteroid signaling to shade avoidance (Djakovic-Petrovic et al. 2007; Bai et al. 2012b; Gallego-Bartolome et al. 2012). Canopy-light-induced DELLA degradation that is mediated through PIF transcription factors appears to be a prerequisite for shade-avoidance responses. PIF4 physically interacts with transcription factor BRASSINAZOLE-RESISTANT 1 (BZR1) to modulate the brassinosteroid responses for elongation growth under shade (Oh et al. 2012).

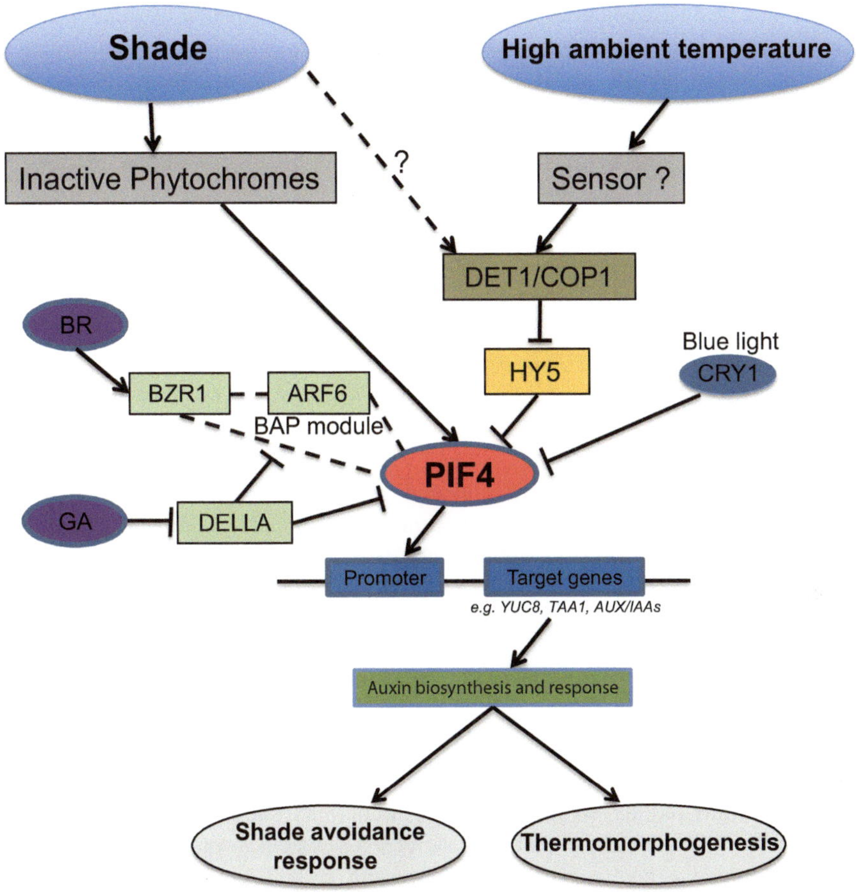

Fig. 8.1 Genetic basis of response of plants to shade and high ambient temperature. PIF4 is the central hub regulating both shade-avoidance and high-temperature response. Downstream to PIF4, auxin biosynthesis and response are critical for the elongation response. Brassinosteroid (BR) and gibberellic acid (GA) signaling regulate PIF4-mediated elongation response via DELLA repressor proteins and BRASSINAZOLE RESISTANT1 (BZR1*)* transcription factor, respectively. The BAP module (BZR1, ARF6, and PIF4; shown by *dotted triangle* in the figure) is controlled by BR and GA signaling to fine-tune the elongation response. Upstream to IF4, shade conditions inactivate phytochromes and thus stabilize PIF4 for the shade-avoidance response. Ambient high temperature increases PIF4 levels through DET1/COP1/HY5 module to exhibit thermomorphogenesis. *Blue* light inhibits temperature-mediated elongation response through direct interaction between CRY1 and PIF4. In summary, light and temperature modulate PIF4 levels antagonistically that leads to increased PIF4 levels under shade and high temperature, resulting in the elongation response. *Solid lines* represent experimentally characterized steps, whereas *dashed lines* indicate potential connections without biological validation. *Pointed-* and *blunt-headed lines* depict activation and suppression, respectively

Moreover, DELLAs negatively regulate brassinosteroid signaling by binding BZR1, thus aiding to the shade-avoidance response (Fig. 8.1) (Gallego-Bartolome et al. 2012). The elongation growth response of the petiole and stem during shade avoidance is also associated with decrease in leaf-blade area (Carabelli et al. 2007). This growth arrest is proposed to be due to an auxin-dependent and a low red/far-red-dependent local decrease in cytokinin concentration that results from upregulation of *CYTOKININ OXIDASE 6* (*CKX6*) in the leaves (Carabelli et al. 2007). Though the significant progress has been made in basic understanding of mechanism underlying the shade-avoidance response, residual responses in multiple *pif* mutants and DELLA-deficient mutants demonstrate the involvement of additional, as yet unidentified, molecular mechanisms.

In addition to increased elongation growth, plants also exhibit reduced branching under shade conditions of dense canopies. *BRANCHED1* (*BRC1*) and *BRC2*, a class II TEOSINTE BRANCHED1, CYCLOIDEA, and PCF [TCP] transcription factors are involved in the shade-induced suppression of branching in *Arabidopsis*. *BRC1* transcription is positively regulated in response to shade (low red/far-red) and is negatively regulated by phyB (Finlayson et al. 2010; Krishna Reddy and Finlayson 2014; Gonzalez-Grandio et al. 2013). Abscisic acid-responsive genes, and cell cycle and ribosome-related genes may act downstream to *BRC1* to control shade-regulated lateral branching (Gonzalez-Grandio et al. 2013). Moreover, plants grown under shade (low red/far-red ratio) flower early (Cerdan and Chory 2003; Halliday et al. 2003). It has been shown that reduction of phyB activity under shade leads to accumulation of CO protein, and subsequently increased FT expression and accelerated flowering (Jang et al. 2008; Valverde et al. 2004).

To summarize, a simple pathway links shade signals to target shade-avoidance genes. The low red/far-red caused by neighbors inactivates photoreceptors, particularly phyB. This results in stabilization of PIFs, which in turn bind and activate auxin-synthesis genes to promote stem growth (Fig. 8.1). Wired to this simple pathway are a complex set of regulatory loops that include links to gibberellins, brassinosteroids, and the circadian clock.

8.5.2.2 Molecular Mechanisms Underlying High Ambient Temperature Response

PIF4 is also a central regulator for the response of plant to higher ambient temperature. PIF4, with a contribution from PIF5, performs pivotal function in high-temperature signaling by manifesting transcriptional responses to trigger hormone-induced developmental changes (Proveniers and van Zanten 2013). Similar to shade-avoidance response, phytohormone biosynthesis and signaling genes are the prominent PIF4 targets, thereby integrating the long-known roles of phytohormones in thermomorphogenesis (Fig. 8.1) (Stavang et al. 2009). Auxin and auxin signaling are necessary and sufficient for PIF4-mediated thermomorphogenic responses (Sun et al. 2012; Franklin et al. 2011). Temperature-mediated binding of PIF4 to the promoters of auxin biosynthesis genes like *YUCCA8 (YUC8)*, *CYTOCHROME P450 FAMILY 79B (CYP79B)*, and *TRYPTOPHAN AMINOTRANSFERASE OF ARABIDOPSIS 1 (TAA1)* increases the levels of indole-3-acetic acid (IAA) at higher ambient temperatures (Franklin et al. 2011; Sun et al. 2012). Consistent with this, PIF4 mutant does not show increase in the levels of auxin and elongation response at high temperature (Franklin et al. 2011; Gray et al. 1998; Sun et al. 2012). Increased level of auxins, then, induces expression of the members of *SMALL AUXIN UPREGULATED RNA* (*SAUR*) gene family to induce hypocotyl elongation at warm temperatures (Spartz et al. 2014). Moreover, high-temperature-induced expression of *EXPANSIN*, a cell-wall-loosening enzyme that directly affects cell elongation, also depends on PIF4 (Bai et al. 2012a). In addition to auxin, BR and GA also regulate elongation growth in response to high temperature as reported for shade conditions (Bai et al. 2012b; Gray et al. 1998; Koini et al. 2009; Stavang et al. 2009; Wang et al. 2014). Regulation of temperature-induced hypocotyl elongation in a PIF-dependent manner involves direct interaction of *BRASSINAZOLE RESISTANT1 (BZR1)*, a transcription factor in BR signaling pathway, with PIF4 (Oh et al. 2012). PIF4 and BZR1 directly interact with AUXIN RESPONSE FACTOR 6 (ARF6), and the resulting *BZR1/ARF6/PIF4* (BAP) module regulates downstream genes, such as auxin biosynthetic and response genes, to trigger elongation growth (Wang et al. 2014). Elevated ambient temperature results in rapid upregulation of major GA biosynthesis genes *AtGA20ox1* and *AtGA3ox1* in *Arabidopsis*, whereas downregulation of GA catabolism genes that results into increased levels of GA that inhibits DELLA proteins to induce PIF4-mediated thermomorphogenic responses (Djakovic-Petrovic et al. 2007; Bai et al. 2012b). In addition, elevated levels of GA release DELLA-mediated repression of BZR1 and ARF6 to allow BAP-module function and subsequent induction of hypocotyl elongation.

Thus, the elongation growth in response to both shade and high-temperature conditions involves PIF4-mediated increase in auxin biosynthesis that results in increased auxin levels (Fig. 8.1). Further GA and BR signaling integrates to this PIF4-mediated cascade to result in elongation growth. Under the shade condition, modulation of phytochrome activity that results from low red/far-red ratio is the major driving force for the activation of PIF4. However, signaling components upstream to PIF4 to induce temperature-mediated response are largely uncharacterized. One such mechanism to induce the transcription of PIF4 in response to high temperature may involve eviction of H2A.Z nucleosomes from PIF promoters at higher temperature (Kumar and Wigge 2010; Wigge 2013). A recent study has shown that DE-ETIOLATED 1 (DET1)-CONSTITUTIVE PHOTOMORPHOGENIC 1 (COP1)-ELONGATED HYPOCOTYL 5 (HY5) module, an integral component of light signaling mechanism, also regulates PIF4 in temperature-dependent manner to induce the elongation response (Delker et al. 2014). Consistent with this, *det1*, along with other photomorphogenic mutants, displays alteration in temperature-induced hypocotyl elongation (Delker et al. 2014). In addition, photoreceptors, cryptochrome 1, physically interact with PIF4 to regulate thermomorphogenesis in blue-light-dependent manner (Ma et al. 2016). Taken together, these studies show that two of the most important environmental factors, light and temperature, share a much larger set of signaling components beyond PIF4 to translate these environmental stimuli to adaptational growth processes.

8.6 Perspective and Future Directions

Light and temperature signals provide plants with the capacity to exquisitely monitor their ambient environmental conditions and respond to changes in surroundings. Based on the literature information available to date, an integrated picture of complex cross talk of light and temperature signaling, with PIF4 as central hub and shared phytohormone pathways, can be comprehended (Fig. 8.1). Further dissection of such signal cross talk should, likely, provide a more comprehensive understanding of how plants develop in fluctuating natural environments. Moreover, these interaction studies will also highlight the evolutionarily advantage of such interconnected systems over the signaling in isolation as the interaction of environmental signaling pathways via shared regulators may enable plants to survive under multiple stress situations.

Rapid increase in world population and associated reduction in agricultural land in conjunction with global climatic changes are one of the major concerns for food security and sustainable growth and development. Supraoptimal temperature and light limitation through shading by neighboring vegetation are among the most damaging abiotic factors to plant survival in natural environments, causing considerable reduction in plant biomass and raising concerns over future crop productivity. Indeed, small (1.5 °C) increases in temperature are predicted to have considerably adverse effects on crop yields. Molecular dissection of these signaling mechanisms and their interaction in crops will certainly help to breed cultivars resistant to unfavorable light and temperature conditions via more precise molecular design approaches.

Acknowledgments Projects on "understanding interaction of light and temperature signaling" at AR laboratory are supported by National Institute of Plant Genome Research core funding and DBT-Ramalingaswami reentry fellowship grant (BT/RLF/reentry/05/2013). JB is supported by research fellowship from Council of Scientific and Industrial Research, India.

References

Ahmad M, Jarillo JA, Smirnova O, Cashmore AR (1998) The CRY1 blue light photoreceptor of Arabidopsis interacts with phytochrome A in vitro. Mol Cell 1(7):939–948

Alabadi D, Blazquez MA (2009) Molecular interactions between light and hormone signaling to control plant growth. Plant Mol Biol 69(4):409–417. doi:10.1007/s11103-008-9400-y

Al-Sady B, Ni W, Kircher S, Schafer E, Quail PH (2006) Photoactivated phytochrome induces rapid PIF3 phosphorylation prior to proteasome-mediated degradation. Mol Cell 23(3):439–446. doi:10.1016/j.molcel.2006.06.011

Araujo WL, Fernie AR, Nunes-Nesi A (2011) Control of stomatal aperture: a renaissance of the old guard. Plant Signal Behav 6(9):1305–1311

Asada K (1996) Radical production and scavenging in the chloroplasts. In: Baker NR (ed) Photosynthesis and the environment. Kluwer Academic Publishers, Dordrecht, pp 123–150

Atkin OK, Loveys BR, Atkinson LJ, Pons TL (2006) Phenotypic plasticity and growth temperature: understanding interspecific variability. J Exp Bot 57(2):267–281. doi:10.1093/jxb/erj029

Auge GA, Rugnone ML, Cortes LE, Gonzalez CV, Zarlavsky G, Boccalandro HE, Sanchez RA (2012) Phytochrome A increases tolerance to high evaporative demand. Physiol Plant 146(2):228–235. doi:10.1111/j.1399-3054.2012.01625.x

Bai MY, Fan M, Oh E, Wang ZY (2012a) A triple helix-loop-helix/basic helix-loop-helix cascade controls cell elongation downstream of multiple hormonal and environmental signaling pathways in Arabidopsis. Plant Cell 24(12):4917–4929. doi:10.1105/tpc.112.105163

Bai MY, Shang JX, Oh E, Fan M, Bai Y, Zentella R, Sun TP, Wang ZY (2012b) Brassinosteroid, gibberellin and phytochrome impinge on a common transcription module in Arabidopsis. Nat Cell Biol 14(8):810–817. doi:10.1038/ncb2546

Baker NR, Harbinson J, Kramer DM (2007) Determining the limitations and regulation of photosynthetic energy transduction in leaves. Plant Cell Environ 30(9):1107–1125. doi:10.1111/j.1365-3040.2007.01680.x

Balasubramanian S, Sureshkumar S, Lempe J, Weigel D (2006) Potent induction of *Arabidopsis thaliana* flowering by elevated growth temperature. PLoS Genet 2(7):e106. doi:10.1371/journal.pgen.0020106

Ball L, Accotto GP, Bechtold U, Creissen G, Funck D, Jimenez A, Kular B, Leyland N, Mejia-Carranza J, Reynolds H, Karpinski S, Mullineaux PM (2004) Evidence for a direct link between glutathione biosynthesis and stress defense gene expression in Arabidopsis. Plant Cell 16(9):2448–2462. doi:10.1105/tpc.104.022608

Bauer D, Viczian A, Kircher S, Nobis T, Nitschke R, Kunkel T, Panigrahi KC, Adam E, Fejes E, Schafer E, Nagy F (2004) Constitutive photomorphogenesis 1 and multiple photoreceptors control degradation of phytochrome interacting factor 3, a transcription factor required for light signaling in Arabidopsis. Plant Cell 16(6):1433–1445. doi:10.1105/tpc.021568

Bita CE, Gerats T (2013) Plant tolerance to high temperature in a changing environment: scientific fundamentals and production of heat stress-tolerant crops. Front Plant Sci 4:273. doi:10.3389/fpls.2013.00273

Blazquez MA, Ahn JH, Weigel D (2003) A thermosensory pathway controlling flowering time in *Arabidopsis thaliana*. Nat Genet 33(2):168–171. doi:10.1038/ng1085

Boccalandro HE, Giordano CV, Ploschuk EL, Piccoli PN, Bottini R, Casal JJ (2012) Phototropins but not cryptochromes mediate the blue light-specific promotion of stomatal conductance, while both enhance photosynthesis and transpiration under full sunlight. Plant Physiol 158(3):1475–1484. doi:10.1104/pp.111.187237

Box MS, Huang BE, Domijan M, Jaeger KE, Khattak AK, Yoo SJ, Sedivy EL, Jones DM, Hearn TJ, Webb AA, Grant A, Locke JC, Wigge PA (2015) ELF3 controls thermoresponsive growth in Arabidopsis. Curr Biol 25(2):194–199. doi:10.1016/j.cub.2014.10.076

Carabelli M, Morelli G, Whitelam G, Ruberti I (1996) Twilight-zone and canopy shade induction of the Athb-2 homeobox gene in green plants. Proc Natl Acad Sci U S A 93(8): 3530–3535

Carabelli M, Possenti M, Sessa G, Ciolfi A, Sassi M, Morelli G, Ruberti I (2007) Canopy shade causes a rapid and transient arrest in leaf development through auxin-induced cytokinin oxidase activity. Genes Dev 21(15):1863–1868. doi:10.1101/gad.432607

Carpaneto A, Ivashikina N, Levchenko V, Krol E, Jeworutzki E, Zhu JK, Hedrich R (2007) Cold transiently activates calcium-permeable channels in Arabidopsis mesophyll cells. Plant Physiol 143(1):487–494. doi:10.1104/pp.106.090928

Carriedo LG, Maloof JN, Brady SM (2016) Molecular control of crop shade avoidance. Curr Opin Plant Biol 30:151–158. doi:10.1016/j.pbi.2016.03.005

Casal JJ (2012) Shade avoidance. Arabidopsis Book 10:e0157. doi:10.1199/tab.0157

Casal JJ (2013) Photoreceptor signaling networks in plant responses to shade. Annu Rev Plant Biol 64:403–427. doi:10.1146/annurev-arplant-050312-120221

Casal JJ, Sanchez RA (1998) Phytochromes and seed germination. Seed Sci Res 8:317–329

Casal JJ, Ballare CL, Tourn M, Sanchez RA (1994) Anatomy, growth and survival of a long-hypocotyl mutant of *Cucumis sativus* deficient in phytochrome B. Ann Bot 73:569–575

Castillon A, Shen H, Huq E (2007) Phytochrome Interacting Factors: central players in phytochrome-mediated light signaling networks. Trends Plant Sci 12(11):514–521. doi:10.1016/j.tplants.2007.10.001

Cerdan PD, Chory J (2003) Regulation of flowering time by light quality. Nature 423(6942):881–885. doi:10.1038/nature01636

Chory J, Peto C, Feinbaum R, Pratt L, Ausubel F (1989) *Arabidopsis thaliana* mutant that develops as a light-grown plant in the absence of light. Cell 58(5):991–999

Christie JM (2007) Phototropin blue-light receptors. Annu Rev Plant Biol 58:21–45. doi:10.1146/annurev.arplant.58.032806.103951

Crawford AJ, McLachlan DH, Hetherington AM, Franklin KA (2012) High temperature exposure increases plant cooling capacity. Curr Biol 22(10):R396–R397. doi:10.1016/j.cub.2012.03.044

Dafny-Yelin M, Tzfira T, Vainstein A, Adam Z (2008) Non-redundant functions of sHSP-CIs in acquired thermotolerance and their role in early seed development in Arabidopsis. Plant Mol Biol 67(4):363–373. doi:10.1007/s11103-008-9326-4

de Lucas M, Prat S (2014) PIFs get BRright: PHYTOCHROME INTERACTING FACTORs as integrators of light and hormonal signals. New Phytol 202(4):1126–1141. doi:10.1111/nph.12725

Delker C, Sonntag L, James GV, Janitza P, Ibanez C, Ziermann H, Peterson T, Denk K, Mull S, Ziegler J, Davis SJ, Schneeberger K, Quint M (2014) The DET1-COP1-HY5 pathway constitutes a multipurpose signaling module regulating plant photomorphogenesis and thermomorphogenesis. Cell Rep 9(6):1983–1989. doi:10.1016/j.celrep.2014.11.043

Deng XW, Matsui M, Wei N, Wagner D, Chu AM, Feldmann KA, Quail PH (1992) COP1, an Arabidopsis regulatory gene, encodes a protein with both a zinc-binding motif and a G beta homologous domain. Cell 71(5):791–801

Devlin PF, Patel SR, Whitelam GC (1998) Phytochrome E influences internode elongation and flowering time in Arabidopsis. Plant Cell 10(9):1479–1487

Djakovic-Petrovic T, de Wit M, Voesenek LA, Pierik R (2007) DELLA protein function in growth responses to canopy signals. Plant J 51(1):117–126. doi:10.1111/j.1365-313X.2007.03122.x

Donohue K, Heschel MS, Chiang GC, Butler CM, Barua D (2007) Phytochrome mediates germination responses to multiple seasonal cues. Plant Cell Environ 30(2):202–212. doi:10.1111/j.1365-3040.2006.01619.x

Duek PD, Fankhauser C (2005) bHLH class transcription factors take centre stage in phytochrome signalling. Trends Plant Sci 10(2):51–54. doi:10.1016/j.tplants.2004.12.005

Duek PD, Elmer MV, van Oosten VR, Fankhauser C (2004) The degradation of HFR1, a putative bHLH class transcription factor involved in light signaling, is regulated by phosphorylation and requires COP1. Curr Biol 14(24):2296–2301. doi:10.1016/j.cub.2004.12.026

Eberhard S, Finazzi G, Wollman FA (2008) The dynamics of photosynthesis. Annu Rev Genet 42:463–515. doi:10.1146/annurev.genet.42.110807.091452

Farquhar GD, Sharkey TD (1982) Stomatal conductance and photosynthesis. Annu Rev Plant Physiol 33:317–345

Finka A, Cuendet AF, Maathuis FJ, Saidi Y, Goloubinoff P (2012) Plasma membrane cyclic nucleotide gated calcium channels control land plant thermal sensing and acquired thermotolerance. Plant Cell 24(8):3333–3348. doi:10.1105/tpc.112.095844

Finlayson SA, Krishnareddy SR, Kebrom TH, Casal JJ (2010) Phytochrome regulation of branching in Arabidopsis. Plant Physiol 152(4):1914–1927. doi:10.1104/pp.109.148833

Franklin KA (2008) Shade avoidance. New Phytol 179(4):930–944. doi:10.1111/j.1469-8137.2008.02507.x

Franklin KA, Quail PH (2010) Phytochrome functions in Arabidopsis development. J Exp Bot 61(1):11–24. doi:10.1093/jxb/erp304

Franklin KA, Whitelam GC (2004) Light signals, phytochromes and cross-talk with other environmental cues. J Exp Bot 55(395):271–276. doi:10.1093/jxb/erh026

Franklin KA, Whitelam GC (2007) Light-quality regulation of freezing tolerance in *Arabidopsis thaliana*. Nat Genet 39(11):1410–1413. doi:10.1038/ng.2007.3

Franklin KA, Praekelt U, Stoddart WM, Billingham OE, Halliday KJ, Whitelam GC (2003) Phytochromes B, D, and E act redundantly to control multiple physiological responses in Arabidopsis. Plant Physiol 131(3):1340–1346. doi:10.1104/pp.102.015487

Franklin KA, Larner VS, Whitelam GC (2005) The signal transducing photoreceptors of plants. Int J Dev Biol 49(5–6):653–664. doi:10.1387/ijdb.051989kf

Franklin KA, Lee SH, Patel D, Kumar SV, Spartz AK, Gu C, Ye S, Yu P, Breen G, Cohen JD, Wigge PA, Gray WM (2011) Phytochrome-interacting factor 4 (PIF4) regulates auxin biosynthesis at high temperature. Proc Natl Acad Sci U S A 108(50):20231–20235. doi:10.1073/pnas.1110682108

Franklin KA, Toledo-Ortiz G, Pyott DE, Halliday KJ (2014) Interaction of light and temperature signalling. J Exp Bot 65(11):2859–2871. doi:10.1093/jxb/eru059

Fryer MJ, Ball L, Oxborough K, Karpinski S, Mullineaux PM, Baker NR (2003) Control of Ascorbate Peroxidase 2 expression by hydrogen peroxide and leaf water status during excess light stress reveals a functional organisation of Arabidopsis leaves. Plant J 33(4):691–705

Gallego-Bartolome J, Minguet EG, Grau-Enguix F, Abbas M, Locascio A, Thomas SG, Alabadi D, Blazquez MA (2012) Molecular mechanism for the interaction between gibberellin and brassinosteroid signaling pathways in Arabidopsis. Proc Natl Acad Sci U S A 109(33):13446–13451. doi:10.1073/pnas.1119992109

Galvez-Valdivieso G, Fryer MJ, Lawson T, Slattery K, Truman W, Smirnoff N, Asami T, Davies WJ, Jones AM, Baker NR, Mullineaux PM (2009) The high light response in Arabidopsis involves ABA signaling between vascular and bundle sheath cells. Plant Cell 21(7):2143–2162. doi:10.1105/tpc.108.061507

Gonzalez-Grandio E, Poza-Carrion C, Sorzano CO, Cubas P (2013) BRANCHED1 promotes axillary bud dormancy in response to shade in Arabidopsis. Plant Cell 25(3):834–850. doi:10.1105/tpc.112.108480

Gray WM, Ostin A, Sandberg G, Romano CP, Estelle M (1998) High temperature promotes auxin-mediated hypocotyl elongation in Arabidopsis. Proc Natl Acad Sci U S A 95(12):7197–7202

Halliday KJ, Whitelam GC (2003) Changes in photoperiod or temperature alter the functional relationships between phytochromes and reveal roles for phyD and phyE. Plant Physiol 131(4):1913–1920. doi:10.1104/pp.102.018135

Halliday KJ, Salter MG, Thingnaes E, Whitelam GC (2003) Phytochrome control of flowering is temperature sensitive and correlates with expression of the floral integrator FT. Plant J 33(5):875–885

Han Y, Song P, Kim J (2007) Phytochrome-Mediated Photomorphogenesis in Plants. J Plant Biol 50(3):230–240

Hasanuzzaman M, Nahar K, Alam MM, Roychowdhury R, Fujita M (2013) Physiological, biochemical, and molecular mechanisms of heat stress tolerance in plants. Int J Mol Sci 14(5):9643–9684. doi:10.3390/ijms14059643

Heggie L, Halliday KJ (2005) The highs and lows of plant life: temperature and light interactions in development. Int J Dev Biol 49(5–6):675–687. doi:10.1387/ijdb.041926lh

Henderson IR, Dean C (2004) Control of Arabidopsis flowering: the chill before the bloom. Development 131(16):3829–3838. doi:10.1242/dev.01294

Hennig L, Stoddart WM, Dieterle M, Whitelam GC, Schafer E (2002) Phytochrome E controls light-induced germination of Arabidopsis. Plant Physiol 128(1):194–200

Henriques R, Jang IC, Chua NH (2009) Regulated proteolysis in light-related signaling pathways. Curr Opin Plant Biol 12(1):49–56. doi:10.1016/j.pbi.2008.10.009

Heschel MS, Selby J, Butler C, Whitelam GC, Sharrock RA, Donohue K (2007) A new role for phytochromes in temperature-dependent germination. New Phytol 174(4):735–741. doi:10.1111/j.1469-8137.2007.02044.x

Hoecker U (2005) Regulated proteolysis in light signaling. Curr Opin Plant Biol 8(5):469–476. doi:10.1016/j.pbi.2005.07.002

Hoecker U, Quail PH (2001) The phytochrome A-specific signaling intermediate SPA1 interacts directly with COP1, a constitutive repressor of light signaling in Arabidopsis. J Biol Chem 276(41):38173–38178. doi:10.1074/jbc.M103140200

Hornitschek P, Kohnen MV, Lorrain S, Rougemont J, Ljung K, Lopez-Vidriero I, Franco-Zorrilla JM, Solano R, Trevisan M, Pradervand S, Xenarios I, Fankhauser C (2012) Phytochrome interacting factors 4 and 5 control seedling growth in changing light conditions by directly controlling auxin signaling. Plant J 71(5):699–711. doi:10.1111/j.1365-313X.2012.05033.x

Imaizumi T, Kay SA (2006) Photoperiodic control of flowering: not only by coincidence. Trends Plant Sci 11(11):550–558. doi:10.1016/j.tplants.2006.09.004

Ishikawa M, Kiba T, Chua NH (2006) The Arabidopsis SPA1 gene is required for circadian clock function and photoperiodic flowering. Plant J 46(5):736–746. doi:10.1111/j.1365-313X.2006.02737.x

Jang S, Marchal V, Panigrahi KC, Wenkel S, Soppe W, Deng XW, Valverde F, Coupland G (2008) Arabidopsis COP1 shapes the temporal pattern of CO accumulation conferring a photoperiodic flowering response. EMBO J 27(8):1277–1288. doi:10.1038/emboj.2008.68

Jiao Y, Lau OS, Deng XW (2007) Light-regulated transcriptional networks in higher plants. Nat Rev Genet 8(3):217–230. doi:10.1038/nrg2049

Jin B, Wang L, Wang J, Jiang KZ, Wang Y, Jiang XX, Ni CY, Wang YL, Teng NJ (2011) The effect of experimental warming on leaf functional traits, leaf structure and leaf biochemistry in *Arabidopsis thaliana*. BMC Plant Biol 11:35. doi:10.1186/1471-2229-11-35

Kami C, Lorrain S, Hornitschek P, Fankhauser C (2010) Light-regulated plant growth and development. Curr Top Dev Biol 91:29–66. doi:10.1016/S0070-2153(10)91002-8

Karpinski S, Reynolds H, Karpinska B, Wingsle G, Creissen G, Mullineaux P (1999) Systemic signaling and acclimation in response to excess excitation energy in Arabidopsis. Science 284(5414):654–657

Kendrick RE, Kronenberg GHM (1994) Photomorphogenesis in land plants, 2nd edn. Kluwer Academic, Dordrecht

Kim TH, Kim BH, von Arnim AG (2002) Repressors of photomorphogenesis. Int Rev Cytol 220:185–223

Kim WY, Fujiwara S, Suh SS, Kim J, Kim Y, Han L, David K, Putterill J, Nam HG, Somers DE (2007) ZEITLUPE is a circadian photoreceptor stabilized by GIGANTEA in blue light. Nature 449(7160):356–360. doi:10.1038/nature06132

Kinoshita T, Doi M, Suetsugu N, Kagawa T, Wada M, Shimazaki K (2001) Phot1 and phot2 mediate blue light regulation of stomatal opening. Nature 414(6864):656–660. doi:10.1038/414656a

Kircher S, Gil P, Kozma-Bognar L, Fejes E, Speth V, Husselstein-Muller T, Bauer D, Adam E, Schafer E, Nagy F (2002) Nucleocytoplasmic partitioning of the plant photoreceptors phytochrome A, B, C, D, and E is regulated differentially by light and exhibits a diurnal rhythm. Plant Cell 14(7):1541–1555

Koini MA, Alvey L, Allen T, Tilley CA, Harberd NP, Whitelam GC, Franklin KA (2009) High temperature-mediated adaptations in plant architecture require the bHLH transcription factor PIF4. Curr Biol 19(5):408–413. doi:10.1016/j.cub.2009.01.046

Krishna Reddy S, Finlayson SA (2014) Phytochrome B promotes branching in Arabidopsis by suppressing auxin signaling. Plant Physiol 164(3):1542–1550. doi:10.1104/pp.113.234021

Kumar SV, Wigge PA (2010) H2A.Z-containing nucleosomes mediate the thermosensory response in Arabidopsis. Cell 140(1):136–147. doi:10.1016/j.cell.2009.11.006

Kumar SV, Lucyshyn D, Jaeger KE, Alos E, Alvey E, Harberd NP, Wigge PA (2012) Transcription factor PIF4 controls the thermosensory activation of flowering. Nature 484(7393):242–245. doi:10.1038/nature10928

Laubinger S, Fittinghoff K, Hoecker U (2004) The SPA quartet: a family of WD-repeat proteins with a central role in suppression of photomorphogenesis in arabidopsis. Plant Cell 16(9):2293–2306. doi:10.1105/tpc.104.024216

Laubinger S, Marchal V, Le Gourrierec J, Wenkel S, Adrian J, Jang S, Kulajta C, Braun H, Coupland G, Hoecker U (2006) Arabidopsis SPA proteins regulate photoperiodic flowering and interact with the floral inducer CONSTANS to regulate its stability. Development 133(16):3213–3222. doi:10.1242/dev.02481

Lee JH, Yoo SJ, Park SH, Hwang I, Lee JS, Ahn JH (2007) Role of SVP in the control of flowering time by ambient temperature in Arabidopsis. Genes Dev 21(4):397–402. doi:10.1101/gad.1518407

Lee HJ, Jung JH, Cortes Llorca L, Kim SG, Lee S, Baldwin IT, Park CM (2014) FCA mediates thermal adaptation of stem growth by attenuating auxin action in Arabidopsis. Nat Commun 5:5473. doi:10.1038/ncomms6473

Leivar P, Monte E (2014) PIFs: systems integrators in plant development. Plant Cell 26(1):56–78. doi:10.1105/tpc.113.120857

Leivar P, Quail PH (2011) PIFs: pivotal components in a cellular signaling hub. Trends Plant Sci 16(1):19–28. doi:10.1016/j.tplants.2010.08.003

Li QH, Yang HQ (2007) Cryptochrome signaling in plants. Photochem Photobiol 83(1):94–101. doi:10.1562/2006-02-28-IR-826

Lin C, Shalitin D (2003) Cryptochrome structure and signal transduction. Annu Rev Plant Biol 54:469–496. doi:10.1146/annurev.arplant.54.110901.160901

Liu H, Yu X, Li K, Klejnot J, Yang H, Lisiero D, Lin C (2008) Photoexcited CRY2 interacts with CIB1 to regulate transcription and floral initiation in Arabidopsis. Science 322(5907):1535–1539. doi:10.1126/science.1163927

Lopez RG, Runkle ES (2004) The effect of temperature on leaf and flower development and flower longevity of Zygopetalum Redvale Fire Kiss' Orchid. Hort Sci 39(7):1630–1634

Lorrain S, Allen T, Duek PD, Whitelam GC, Fankhauser C (2008) Phytochrome-mediated inhibition of shade avoidance involves degradation of growth-promoting bHLH transcription factors. Plant J 53(2):312–323. doi:10.1111/j.1365-313X.2007.03341.x

Ma D, Li X, Guo Y, Chu J, Fang S, Yan C, Noel JP, Liu H (2016) Cryptochrome 1 interacts with PIF4 to regulate high temperature-mediated hypocotyl elongation in response to blue light. Proc Natl Acad Sci U S A 113(1):224–229. doi:10.1073/pnas.1511437113

Mao J, Zhang YC, Sang Y, Li QH, Yang HQ (2005) From the cover: a role for Arabidopsis cryptochromes and COP1 in the regulation of stomatal opening. Proc Natl Acad Sci U S A 102(34):12270–12275. doi:10.1073/pnas.0501011102

Martinez-Garcia JF, Huq E, Quail PH (2000) Direct targeting of light signals to a promoter element-bound transcription factor. Science 288(5467):859–863

Mas P, Devlin PF, Panda S, Kay SA (2000) Functional interaction of phytochrome B and cryptochrome 2. Nature 408(6809):207–211. doi:10.1038/35041583

McClung CR, Davis SJ (2010) Ambient thermometers in plants: from physiological outputs towards mechanisms of thermal sensing. Curr Biol 20(24):R1086–R1092. doi:10.1016/j.cub.2010.10.035

McCormac AC, Terry MJ (2002) Light-signalling pathways leading to the co-ordinated expression of HEMA1 and Lhcb during chloroplast development in *Arabidopsis thaliana*. Plant J 32(4):549–559

Miller G, Schlauch K, Tam R, Cortes D, Torres MA, Shulaev V, Dangl JL, Mittler R (2009) The plant NADPH oxidase RBOHD mediates rapid systemic signaling in response to diverse stimuli. Sci Signal 2(84):ra45. doi:10.1126/scisignal.2000448

Mishkind M, Vermeer JE, Darwish E, Munnik T (2009) Heat stress activates phospholipase D and triggers PIP accumulation at the plasma membrane and nucleus. Plant J 60(1):10–21. doi:10.1111/j.1365-313X.2009.03933.x

Mittler R, Vanderauwera S, Gollery M, Van Breusegem F (2004) Reactive oxygen gene network of plants. Trends Plant Sci 9(10):490–498. doi:10.1016/j.tplants.2004.08.009

Mittler R, Finka A, Goloubinoff P (2012) How do plants feel the heat? Trends Biochem Sci 37(3):118–125. doi:10.1016/j.tibs.2011.11.007

Miyazaki Y, Takase T, Kiyosue T (2015) ZEITLUPE positively regulates hypocotyl elongation at warm temperature under light in *Arabidopsis thaliana*. Plant Signal Behav 10(5):e998540. doi:10.1080/15592324.2014.998540

Mockler T, Yang H, Yu X, Parikh D, Cheng YC, Dolan S, Lin C (2003) Regulation of photoperiodic flowering by Arabidopsis photoreceptors. Proc Natl Acad Sci U S A 100(4):2140–2145. doi:10.1073/pnas.0437826100

Moglich A, Yang X, Ayers RA, Moffat K (2010) Structure and function of plant photoreceptors. Annu Rev Plant Biol 61:21–47. doi:10.1146/annurev-arplant-042809-112259

Morelli G, Ruberti I (2002) Light and shade in the photocontrol of Arabidopsis growth. Trends Plant Sci 7(9):399–404

Muller P, Li XP, Niyogi KK (2001) Non-photochemical quenching. A response to excess light energy. Plant Physiol 125(4):1558–1566

Nagatani A (2004) Light-regulated nuclear localization of phytochromes. Curr Opin Plant Biol 7(6):708–711. doi:10.1016/j.pbi.2004.09.010

Nagatani A, Kay SA, Deak M, Chua NH, Furuya M (1991) Rice type I phytochrome regulates hypocotyl elongation in transgenic tobacco seedlings. Proc Natl Acad Sci U S A 88(12):5207–5211

Nagatani A, Reed JW, Chory J (1993) Isolation and initial characterization of Arabidopsis mutants that are deficient in phytochrome A. Plant Physiol 102(1):269–277

Niyogi KK (1999) Photoprotection Revisited: genetic and molecular approaches. Annu Rev Plant Physiol Plant Mol Biol 50:333–359. doi:10.1146/annurev.arplant.50.1.333

Oh E, Zhu JY, Wang ZY (2012) Interaction between BZR1 and PIF4 integrates brassinosteroid and environmental responses. Nat Cell Biol 14(8):802–809. doi:10.1038/ncb2545

Ort DR (2001) When there is too much light. Plant Physiol 125(1):29–32

Osterlund MT, Hardtke CS, Wei N, Deng XW (2000) Targeted destabilization of HY5 during light-regulated development of Arabidopsis. Nature 405(6785):462–466. doi:10.1038/35013076

Pedmale UV, Huang SS, Zander M, Cole BJ, Hetzel J, Ljung K, Reis PA, Sridevi P, Nito K, Nery JR, Ecker JR, Chory J (2016) Cryptochromes interact directly with PIFs to control plant growth in limiting blue light. Cell 164(1–2):233–245. doi:10.1016/j.cell.2015.12.018

Penfield S (2008) Temperature perception and signal transduction in plants. New Phytol 179(3):615–628. doi:10.1111/j.1469-8137.2008.02478.x

Procko C, Crenshaw CM, Ljung K, Noel JP, Chory J (2014) Cotyledon-generated auxin is required for shade-induced hypocotyl growth in *Brassica rapa*. Plant Physiol 165(3):1285–1301. doi:10.1104/pp.114.241844

Proveniers MC, van Zanten M (2013) High temperature acclimation through PIF4 signaling. Trends Plant Sci 18(2):59–64. doi:10.1016/j.tplants.2012.09.002

Queitsch C, Hong SW, Vierling E, Lindquist S (2000) Heat shock protein 101 plays a crucial role in thermotolerance in Arabidopsis. Plant Cell 12(4):479–492

Raschke A, Ibanez C, Ullrich KK, Anwer MU, Becker S, Glockner A, Trenner J, Denk K, Saal B, Sun X, Ni M, Davis SJ, Delker C, Quint M (2015) Natural variants of ELF3 affect thermomorphogenesis by transcriptionally modulating PIF4-dependent auxin response genes. BMC Plant Biol 15:197. doi:10.1186/s12870-015-0566-6

Reed JW, Nagpal P, Poole DS, Furuya M, Chory J (1993) Mutations in the gene for the red/far-red light receptor phytochrome B alter cell elongation and physiological responses throughout Arabidopsis development. Plant Cell 5(2):147–157. doi:10.1105/tpc.5.2.147

Reeves PH, Coupland G (2000) Response of plant development to environment: control of flowering by daylength and temperature. Curr Opin Plant Biol 3(1):37–42

Rizzini L, Favory JJ, Cloix C, Faggionato D, O'Hara A, Kaiserli E, Baumeister R, Schafer E, Nagy F, Jenkins GI, Ulm R (2011) Perception of UV-B by the Arabidopsis UVR8 protein. Science 332(6025):103–106. doi:10.1126/science.1200660

Saidi Y, Peter M, Finka A, Cicekli C, Vigh L, Goloubinoff P (2010) Membrane lipid composition affects plant heat sensing and modulates Ca(2+)-dependent heat shock response. Plant Signal Behav 5(12):1530–1533

Salter MG, Franklin KA, Whitelam GC (2003) Gating of the rapid shade-avoidance response by the circadian clock in plants. Nature 426(6967):680–683. doi:10.1038/nature02174

Sasidharan R, Chinnappa CC, Staal M, Elzenga JT, Yokoyama R, Nishitani K, Voesenek LA, Pierik R (2010) Light quality-mediated petiole elongation in Arabidopsis during shade avoidance involves cell wall modification by xyloglucan endotransglucosylase/hydrolases. Plant Physiol 154(2):978–990. doi:10.1104/pp.110.162057

Seo HS, Yang JY, Ishikawa M, Bolle C, Ballesteros ML, Chua NH (2003) LAF1 ubiquitination by COP1 controls photomorphogenesis and is stimulated by SPA1. Nature 423(6943):995–999. doi:10.1038/nature01696

Shen H, Zhu L, Castillon A, Majee M, Downie B, Huq E (2008) Light-induced phosphorylation and degradation of the negative regulator PHYTOCHROME-INTERACTING FACTOR1 from Arabidopsis depend upon its direct physical interactions with photoactivated phytochromes. Plant Cell 20(6):1586–1602. doi:10.1105/tpc.108.060020

Somers DE, Kim WY, Geng R (2004) The F-box protein ZEITLUPE confers dosage-dependent control on the circadian clock, photomorphogenesis, and flowering time. Plant Cell 16(3):769–782. doi:10.1105/tpc.016808

Spartz AK, Ren H, Park MY, Grandt KN, Lee SH, Murphy AS, Sussman MR, Overvoorde PJ, Gray WM (2014) SAUR inhibition of PP2C-D phosphatases activates plasma membrane H+−ATPases to promote cell expansion in Arabidopsis. Plant Cell 26(5):2129–2142. doi:10.1105/tpc.114.126037

Stavang JA, Gallego-Bartolome J, Gomez MD, Yoshida S, Asami T, Olsen JE, Garcia-Martinez JL, Alabadi D, Blazquez MA (2009) Hormonal regulation of temperature-induced growth in Arabidopsis. Plant J 60(4):589–601. doi:10.1111/j.1365-313X.2009.03983.x

Su PH, Li HM (2008) Arabidopsis stromal 70-kD heat shock proteins are essential for plant development and important for thermotolerance of germinating seeds. Plant Physiol 146(3):1231–1241. doi:10.1104/pp.107.114496

Sullivan JA, Deng XW (2003) From seed to seed: the role of photoreceptors in Arabidopsis development. Dev Biol 260(2):289–297

Sun J, Qi L, Li Y, Chu J, Li C (2012) PIF4-mediated activation of YUCCA8 expression integrates temperature into the auxin pathway in regulating arabidopsis hypocotyl growth. PLoS Genet 8(3):e1002594. doi:10.1371/journal.pgen.1002594

Suzuki N, Koussevitzky S, Mittler R, Miller G (2012) ROS and redox signalling in the response of plants to abiotic stress. Plant Cell Environ 35(2):259–270. doi:10.1111/j.1365-3040.2011.02336.x

Tilbrook K, Arongaus AB, Binkert M, Heijde M, Yin R, Ulm R (2013) The UVR8 UV-B Photoreceptor: perception, signaling and response. Arabidopsis Book 11:e0164. doi:10.1199/tab.0164

Triantaphylides C, Krischke M, Hoeberichts FA, Ksas B, Gresser G, Havaux M, Van Breusegem F, Mueller MJ (2008) Singlet oxygen is the major reactive oxygen species involved in photooxidative damage to plants. Plant Physiol 148(2):960–968. doi:10.1104/pp.108.125690

Valverde F, Mouradov A, Soppe W, Ravenscroft D, Samach A, Coupland G (2004) Photoreceptor regulation of CONSTANS protein in photoperiodic flowering. Science 303(5660):1003–1006. doi:10.1126/science.1091761

Vasseur F, Pantin F, Vile D (2011) Changes in light intensity reveal a major role for carbon balance in Arabidopsis responses to high temperature. Plant Cell Environ 34(9):1563–1576. doi:10.1111/j.1365-3040.2011.02353.x

Wang H, Wang H (2015) Phytochrome signaling: time to tighten up the loose ends. Mol Plant 8(4):540–551. doi:10.1016/j.molp.2014.11.021

Wang W, Bai MY, Wang ZY (2014) The brassinosteroid signaling network-a paradigm of signal integration. Curr Opin Plant Biol 21:147–153. doi:10.1016/j.pbi.2014.07.012

Whitelam G, Halliday KJ (2007) Light and plant development. Blackwell, Oxford

Wigge PA (2013) Ambient temperature signalling in plants. Curr Opin Plant Biol 16(5):661–666. doi:10.1016/j.pbi.2013.08.004

Wu SH (2014) Gene expression regulation in photomorphogenesis from the perspective of the central dogma. Annu Rev Plant Biol 65:311–333. doi:10.1146/annurev-arplant-050213-040337

Xiao B, Coste B, Mathur J, Patapoutian A (2011) Temperature-dependent STIM1 activation induces Ca(2)+ influx and modulates gene expression. Nat Chem Biol 7(6):351–358. doi:10.1038/nchembio.558

Yamada K, Fukao Y, Hayashi M, Fukazawa M, Suzuki I, Nishimura M (2007) Cytosolic HSP90 regulates the heat shock response that is responsible for heat acclimation in *Arabidopsis thaliana*. J Biol Chem 282(52):37794–37804. doi:10.1074/jbc.M707168200

Yang J, Lin R, Sullivan J, Hoecker U, Liu B, Xu L, Deng XW, Wang H (2005) Light regulates COP1-mediated degradation of HFR1, a transcription factor essential for light signaling in Arabidopsis. Plant Cell 17(3):804–821. doi:10.1105/tpc.104.030205

Yao J, Liu B, Qin F (2011) Modular thermal sensors in temperature-gated transient receptor potential (TRP) channels. Proc Natl Acad Sci U S A 108(27):11109–11114. doi:10.1073/pnas.1105196108

Yi C, Deng XW (2005) COP1 – from plant photomorphogenesis to mammalian tumorigenesis. Trends Cell Biol 15(11):618–625. doi:10.1016/j.tcb.2005.09.007

Zhu D, Maier A, Lee JH, Laubinger S, Saijo Y, Wang H, Qu LJ, Hoecker U, Deng XW (2008) Biochemical characterization of Arabidopsis complexes containing CONSTITUTIVELY PHOTOMORPHOGENIC1 and SUPPRESSOR OF PHYA proteins in light control of plant development. Plant Cell 20(9):2307–2323. doi:10.1105/tpc.107.056580

Zhu W, Ausin I, Seleznev A, Mendez-Vigo B, Pico FX, Sureshkumar S, Sundaramoorthi V, Bulach D, Powell D, Seemann T, Alonso-Blanco C, Balasubramanian S (2015) Natural variation identifies ICARUS1, a universal gene required for cell proliferation and growth at high temperatures in *Arabidopsis thaliana*. PLoS Genet 11(5):e1005085. doi:10.1371/journal.pgen.1005085

Plant Responses to Combined Drought and Pathogen Infection: Current Understanding on the Role of Phytohormones

9

Prachi Pandey and Muthappa Senthil-Kumar

Abstract

Plants under natural conditions encounter a number of abiotic and biotic stresses often being inflicted simultaneously. Plant responses to a stress are governed by intricate network of the hormone signaling pathways. Abscisic acid (ABA) forms the major component of the plant response to drought and cold stress. Salicylic acid (SA), jasmonic acid (JA), and ethylene act as key regulators of plant response to pathogen infection. In fact, the extensive cross talk among the different hormone-mediated signaling pathways determines plant response to a particular stress. A large number of studies focus on hormone signaling under individual drought and pathogen stresses and the cross talk between the two stress responses. However, owing to the relatively few studies on combined drought and pathogen stresses, our understanding of phytohormonal signaling under combined stress is still obscure. Recent studies on combined drought and pathogen infection indicate that plants when simultaneously exposed to the two stresses often exhibit a transcriptional and metabolic response different from that exhibited under single stress conditions. This is also applicable to the phytohormonal signaling. The nature, time, and severity of the two stresses in combination modulate hormonal concentrations as well as the hormone signal transduction pathways involved. In this chapter, we provide a compendious description of the role of the three major hormones, namely, ABA, SA, and JA, in combined drought and pathogen infection. A brief description of the role of auxins, cytokinins, and gibberellins has also been provided. Taking

P. Pandey • M. Senthil-Kumar (✉)
National Institute of Plant Genome Research, 10531, JNU Campus, Aruna Asaf Ali Marg, New Delhi 110 067, India
e-mail: skmuthappa@nipgr.ac.in

M. Senthil-Kumar (eds.), *Plant Tolerance to Individual and Concurrent Stresses*,
DOI 10.1007/978-81-322-3706-8_9

leads from few studies, we have discussed the potential role of hormones in conferring combined drought and pathogen stress tolerance to plants. We also briefly discussed the effect of different "stress elicitors" on hormone signaling.

Keywords
Combined stress • Drought • Pathogen infection • Phytohormonal signaling • Cross talk • Abscisic acid • Salicylic acid • Jasmonic acid • Stress elicitors

9.1 Introduction

Plants under field conditions encounter a number of biotic and abiotic stresses which often occur simultaneously. Recently, it has been proposed that the response of plants to combined stresses is an outcome of complex interactions between molecular and metabolic responses elicited under individual stress (Atkinson et al. 2013; Rasmussen et al. 2013; Prasch and Sonnewald 2013; Suzuki et al. 2014; Ramegowda and Senthil-Kumar 2015). The complex interactions and intensive cross talks make their response to combined stress different from that elicited in response to individual stresses. The interaction is further dependent on the nature, severity, and timing of the two stresses as well as the developmental stage of the plant (Suzuki et al. 2014; Ramegowda and Senthil-Kumar 2015; Pandey et al. 2015). Phytohormonal signaling which is an important component of plants' defense to various biotic and abiotic stresses (Fujita et al. 2006) is modulated under combined stress conditions, and it could be different from the signaling under individual stresses.

Phytohormones act in complex, regulated networks having several points of interactions among their signaling pathways. ABA signaling pathway is majorly activated in response to drought and known to interact with SA and JA signaling, which regulates defense responses both positively (Adie et al. 2007) and negatively (Audenaert et al. 2002; Anderson et al. 2004; Xu et al. 2013) against different pathogens. Similarly, SA, JA, and Et signaling pathways are shown to interact among themselves as well as with ABA signaling (Spoel and Dong 2008; Miura and Tada 2014; Kazan 2015). SA and JA signaling are known to be mutually antagonistic, and the activation of these two pathways is governed by the type of pathogens. SA signaling regulates plant defense response against biotrophs, whereas JA and Et signaling is involved in responses against necrotrophs (Glazebrook 2005). Antagonistic interaction between SA and ABA signaling and a moderately complex interaction between ABA and JA signaling (consisting of both synergistic and antagonistic interactions) have been reported (Pieterse et al. 2012). Synergism between two signaling pathways may lead to enhanced tolerance under the combined stress, and the antagonism might be helpful in prioritizing one pathway over the other (Spoel and Dong 2008). Although several reports are available unraveling the transcriptional responses of plants under combined stress, the role of hormones has not been discussed in detail, so far. In order to fully comprehend the plant responses under combined stress conditions, understanding the phytohormonal cross talk under combined stress is important. This chapter covers the current understanding of the role of hormones under combined drought and pathogen stresses using the insights gained from recent transcriptomic studies performed under combined drought and pathogen infection. We also discuss the scenarios wherein exogenous ABA treatment and pathogen elicitors are used to mimic the combined stress conditions.

9.2 Cross Talk Among Phytohormonal Signaling Pathways Involved in Drought and Pathogen Infection

9.2.1 ABA-SA Cross Talk

ABA plays a role in drought tolerance and also increases the susceptibility of plants to several pathogens by antagonizing SA signaling (Mauch-Mani and Mauch 2005) (Table 9.1). The exogenous application of ABA is known to reduce SA accumulation in plants thereby leading to enhanced disease susceptibility in a number of pathosystems. The antagonism between ABA and SA has also been shown to be utilized by pathogens to combat the plant defense against them. For example, infection of *Arabidopsis thaliana* with *Pseudomonas syringae* pv. tomato DC3000 (causal agent of bacterial speck) leads to an increase in ABA and a suppression in SA levels, resulting in enhanced susceptibility of the plants to the pathogen (de Torres-Zabala et al. 2007). ABA-deficient mutants which also exhibit enhanced SA levels are known to resist pathogen infection. For example, *A. thaliana* ABA-deficient *Arabidopsis* aldehyde oxidase3 *(aao3)* mutant plants challenged with *P. syringae* pv. tomato DC3000 showed decreased ABA levels, increased SA levels, and consequently reduced pathogen multiplication (de Torres-Zabala et al. 2009). These examples suggest the negative cross talk between ABA and SA signaling pathways in case of pathogen infection.

Drought also modulates the endogenous SA concentration in plants. For example, drought stress leads to increase in endogenous levels of SA in *Phillyrea angustifolia* and *Hordeum vulgare* (Munne-Bosch and Penuelas 2003; Bandurska and Stroiński 2005). Modulation of endogenous levels of SA also enhances drought tolerance. For example, SA-accumulating *A. thaliana* mutants like constitutive expressor of PR genes 5 (*cpr5*), accelerated cell death 6 (*acd6*), and activated disease resistance 1 (*adr1*) exhibit increased drought tolerance (Chini et al. 2004; Miura et al. 2013). Thus, the accumulation of endogenous SA levels leading to drought tolerance suggests a positive cross talk between ABA and SA signaling pathways under drought stress. However, the cross talk seems to be tightly regulated as indicated by the fact that higher concentrations of SA decrease drought tolerance (Borsani et al. 2001; Korkmaz et al. 2007). SA is known to enhance drought tolerance by strengthening the antioxidant defense of plants. For instance, SA pretreatment to *Zea mays* before exposure to drought enhanced the activity of antioxidant enzymes like superoxide dismutase (SOD), ascorbate peroxidase (APX), glutathione reductase (GR), and monodehydroascorbate reductase (MDHAR) and thereby enhanced

Table 9.1 Role of ABA in modulating resistance or susceptibility of plants to various pathogens

S.No.	Plant	Pathogen	Lifestyle of pathogen	Effect of ABA on plant-pathogen interaction	Mechanism/pathway involved	References
1	*Arabidopsis thaliana*	*Ralstonia solanacearum*	Hemibiotroph	Resistance	Activation of ABA signaling leading to expression of genes encoding antibacterial peptides	Hernández-Blanco et al. (2007) and Feng et al. (2012)
2	*A. thaliana*	*Alternaria brassicicola*	Necrotroph	Resistance	ABA-mediated callose deposition	Ton et al. (2009)
3	*A. thaliana*	*Pseudomonas syringae pv. tomato*	Biotroph	Susceptibility	Suppression of SA signaling	Mohr and Cahill (2003, 2007)
4	*A. thaliana*	*Plectosphaerella cucumerina*	Necrotroph	Susceptibility	Suppression of SA, JA/ethylene signaling	Sanchez-Vallet et al. (2012)
5	*Oryza sativa*	*Cochliobolus miyabeanus*	Necrotroph	Resistance	MAP kinase-mediated repression of ethylene signaling	de Vleesschauwer et al. (2010)
6	*O. sativa*	*Xanthomonas oryzae*	Biotroph	Susceptibility	Suppression of SA signaling	Xu et al. (2013)
7	*O. sativa*	*Magnaporthe grisea*	Hemibiotroph	Susceptibility	Suppression of SA signaling	Koga et al. (2004) and Jiang et al. (2010)
8	*Hordeum vulgare*	*Blumeria graminis*	Biotroph	Resistance	Activation of ABA signaling	Wiese et al. (2004)
9	*H. vulgare*	*Magnaporthe oryzae*	Hemi biotroph	Susceptibility	Reduced plant resistance to penetration by fungus	Ulferts et al. (2015)
10	*Solanum lycopersicum*	*Botrytis cinerea*	Necrotroph	Susceptibility	Suppression of SA signaling	Audenaert et al. (2002)

drought tolerance (Saruhan et al. 2012). The drought-tolerant SA-accumulating *A. thaliana* mutants *cpr5* and *acd6* also exhibit stomatal closure (Okuma et al. 2014). SA-accumulating mutants are known to be resistant to pathogens. For instance, the mutant *cpr5* is resistant to *Peronospora parasitica* and *P. syringae* pv. maculicola (Bowling et al. 1997). Furthermore, *adr1* plants are resistant to *P. parasitica* and *Erysiphe cichoracearum* (Grant et al. 2003). Taken together, drought stress results in a positive cross talk between ABA and SA signaling pathways, and under pathogen infection, SA and ABA signaling antagonizes each other.

9.2.1.1 Key Players of ABA-SA Cross Talk

Evidences indicate that the highly regulated cross communication between two signaling pathways is mediated by a number of transcription factors. For instance, overexpression of SA regulatory gene, *Oryza sativa* non-expressor of pathogen-responsive gene 1 (*OsNPR1*), reversed the ABA-mediated susceptibility of *O. sativa* to *Xanthomonas oryzae* indicating NPR1-mediated regulation of SA-ABA cross talk (Xu et al. 2013). Another important protein-mediating SA-ABA cross talk is OsWRKY45 whose expression is upregulated by both drought and pathogen infection. Overexpression of gene encoding OsWRKY45 in *A. thaliana* plants resulted in increased drought tolerance and enhanced resistance to the bacterial pathogen *P. syringae* pv. tomato DC3000 (Qiu and Yu 2009). OsMPK6, responsible for activating WRKY45 through phosphorylation, is also an important node of interaction between the two signaling pathways. Treatment of *O. sativa* plants with ABA leads to dephosphorylation of OsMPK6 by OsPTP1/2, leading to an impaired activity of OsWRKY45 and reduced tolerance to *Magnaporthe oryzae* (Ueno et al. 2015).

9.2.2 ABA-JA Cross Talk

Infection by a pathogen under drought activates either synergistic or antagonistic interaction between JA and ABA signaling pathways depending on the lifestyle (biotroph/necrotroph) of the pathogens. ABA has been shown to modulate wound-induced JA accumulation in *Solanum tuberosum* and *S. lycopersicum* (Peña-Cortés et al. 1995). Moreover, the attenuation of JA accumulation in ABA-deficient *aba2* mutant of *A. thaliana* when challenged with *Pythium irregulare* (causal agent of damping off disease) substantiates the positive modulation of JA signaling by ABA (Adie et al. 2007) ABA-induced JA accumulation also led to resistance of *A. thaliana* to *Alternaria brassicicola* (causal agent of leaf spot) (Fan et al. 2009). Contrastingly, exogenous treatment of ABA led to decrease in resistance of *A. thaliana* plants to the necrotroph, *Fusarium oxysporum* (causal agent of wilt; Anderson et al. 2004; Asselbergh et al. 2008). The contrasting effect of ABA on plant immunity toward different pathogens may be due to the fact that ABA has differential effect on the two branches of JA-Et signaling, namely, the MYC branch that mediates defense against insect herbivores and the ethylene-responsive factor, ERF branch, which responds to defense against necrotrophs. The MYC branch is co-regulated by ABA, and ERF branch is regulated by Et (Lorenzo et al. 2003). While MYC branch regulates the expression of marker gene VSP2 (Lorenzo et al. 2004; Fernández-Calvo et al. 2011), ERF branch regulates the expression of marker gene, plant defensin 1.2 (*PDF1.2*) (Lorenzo et al. 2003; Zhu et al. 2011). ABA and its signaling components positively regulate the MYC branch and negatively regulate the signaling components of ERF branch (Anderson et al. 2004). The ABA-associated susceptibility of *A. thaliana* to *F. oxysporum* can be attributed to AtMYC2-mediated inhibition of expression of *PDF1.2*.

Exogenous application of JA has been shown to impart drought stress tolerance to *Brassica juncea* by enhancing the activities of antioxidant enzymes like GR, MDHAR, and glyoxalase I and to *Pyrus bretschneideri* plants by increasing betaine accumulation (Gao et al. 2004; Alam et al. 2014). Both ABA and MeJA induce stomatal closure, most likely by triggering the production of reactive oxygen species (ROS) in stomatal guard cells (Munemasa et al. 2007). An increase in endogenous level of JA along with ABA was found to occur during the early stages of drought stress in *A. thaliana* (Harb et al. 2010). Drought stress-mediated enhancement in JA levels has also been shown in *O. sativa* (Du et al. 2013). Increase in the concentration of JA and the concomitant increase in ABA levels due to drought stress show its regulatory role on ABA signaling. This is further substantiated by the transient accumulation of JA prior to ABA in case of severe drought stress in roots of citrumelo (*Citrus paradisi* × *Poncirus trifoliata*) cv. CPB 4475 (de Ollas et al. 2013). A recent study using mutants impaired in jasmonate biosynthesis (*opr3, lox6, and jar1–1*) and JA-dependent signaling (*coi1*) also shows that JA accumulation leading to JA-Ile buildup is necessary for the increase in ABA concentrations in roots of *A. thaliana* plants under drought stress (de Ollas et al. 2015). Furthermore, JA-deficient lines exhibit reduced expression of ABA signaling genes like ABA-insensitive 2 (*ABI2*) and ABA-insensitive 5 (*ABI5*) suggesting JA-ABA-positive cross talk (de Ollas et al. 2015).

9.2.2.1 Key Players in ABA-JA Cross Talk

One of the key players in ABA-JA cross talk is AtMYC2 transcription factor which positively regulates ABA signaling pathway by activating the expression of response to dehydration 22 (*RD22*) gene (Abe et al. 2003). AtMYC2 also upregulates the expression of JA-mediated wounding-responsive genes encoding vegetative storage protein (VSP) and lipoxygenase (LOX) and represses the expression of genes under the ERF branch of JA signaling including the genes encoding PR4 and PDF1.2 (Anderson et al. 2004; Lorenzo et al. 2004). Similarly, botrytis-susceptible gene 1 (*BOS1*) acts as a modulator of ABA-JA cross talk by positively regulating ABA and JA signaling (Mengiste et al. 2003). *BOS1* knockdown lines are susceptible to drought and infection by *B. cinerea* and *A. brassicicola* (Melotto et al. 2006).

9.2.3 SA-JA Cross Talk

SA- and JA-dependent defense against biotrophs and necrotrophs, respectively, has been widely reported (Glazebrook 2005; Pieterse et al. 2012). SA-JA cross talk is crucial in governing plants' response when simultaneously attacked by different pathogens and is known to interact either antagonistically or synergistically in response to pathogen infection (Glazebrook 2005; Mur et al. 2006). For instance, induction of SA signaling in *A. thaliana* plants when challenged with avirulent *P. syringae* leads to the suppression of JA signaling by repressing the expression of *PDF1.2* and reduces the susceptibility of plants to the necrotroph *A. brassicicola* (Spoel et al. 2007). In the case of *A. thaliana-P. syringae* interaction, the pathogen produces a virulence factor, coronatine, a JA mimic which suppresses SA signaling in the plants and makes them susceptible to the pathogen (Zheng et al. 2012). Synergistic interactions between SA and JA signaling pathways also occur. For instance, low concentrations of JA and SA lead to increased expression of JA-responsive *PDF1.2* gene and SA-regulated *PR-1* gene (Mur et al. 2006). However, at higher concentrations, the effect becomes antagonistic. For example, JA-induced PDF1.2 expression is found to be enhanced by exogenous application of SA at concentrations of up to 350 μM, but it is reduced at higher levels of SA. Similarly, the expression levels of SA-responsive *PR1* gene in response to 10 μm SA are enhanced by the addition of JA at concentrations up to 125 μm, beyond which *PR1* gene expression was reduced. This indicates that the SA-JA interaction is dependent on the relative concentrations of two hormones (Mur et al. 2006).

9.2.3.1 Key Players in SA-JA Cross Talk

The SA-JA cross talk is known to be modulated by transcription factor NPR1. The role of cytosolic NPR1 in SA-mediated suppression of JA signaling has been shown in *A. thaliana* and *O. sativa* (Spoel et al. 2003). Other important modulators of SA-JA cross talk include TGA transcription factors, suppressor of SA insensitivity 2 (SSI2), and WRKY transcription factors. TGA transcription factors are positive regulators of SA-mediated *PR* gene expression and negatively regulate JA-responsive gene *PDF1.2* (Spoel et al. 2003; Leon-Reyes et al. 2010). Several WRKY transcription factors like WRKY 50, WRKY 51 (Gao et al. 2011), WRKY70 (Li et al. 2006), and WRKY62 (Mao et al. 2007) have been associated with SA-mediated suppression of JA pathway. The role of several other WRKY transcription factors, namely, WRKY8, WRKY 11, WRKY 17, WRKY 18, WRKY 40, WRKY 41, and WRKY 60, has also been indicated in SA-JA cross talk (Journot-Catalino et al. 2006; Xu et al. 2006; Higashi et al. 2008; Chen et al. 2010). Understanding the role of these regulatory proteins in drought stress might help in dissecting the role of SA and JA under combined drought and pathogen stress.

9.2.4 Other Hormones and Their Cross Talk

Auxins, cytokinins (CKs), brassinosteroids (BRs), and gibberellins (GA) are reported to play a role in drought and pathogen stress tolerance (de Vleesschauwer et al. 2014). These hormones are also known to interact with ABA, JA, and SA signaling pathways. For instance, indole acetic acid (IAA) is known to compromise the resistance of plants to biotrophs but enhances tolerance against necrotrophs. Treatment with 1-naphthalacetic acid (NAA) enhances the susceptibility of *A. thaliana* plants to *P. syringae* pv. tomato DC3000 (Chen et al. 2007). The enhanced susceptibility is attributed to loosening of cell wall and increased cell permeability leading to nutrient leakage into apoplast. Moreover, SA signaling antagonizes the IAA signaling and suppresses the expression of small auxin-up RNA (SAUR) family genes and Aux/IAA family genes. SA signaling is also involved in the upregulation of the expression of IAA-amido synthase gene belonging to glycoside hydrolase 3 (GH3) family (Fu and Wang 2011). Notably, overexpression of OsGH3.13 enhances the expression of several late embryogenesis genes (*LEA*) like *OsLEA1*, *OsLEA8a*, *OsLEA14a*, and *OsLEA18* and improves drought tolerance in *O. sativa* (Zhang et al. 2009). In some instances, auxins act independent of SA and JA to modulate plant defense against pathogens. IAA-induced production of expansins makes *O. sativa* vulnerable to *Xanthomonas oryzae* (causal agent of bacterial blight) and *Magnaporthe grisea* (causal agent of blast) (Fu et al. 2011).

Emerging evidences indicate a role of CKs in both abiotic and biotic stress tolerance mechanisms of plants (O'Brien and Benková 2013). Enhanced CK levels are known to correspond to increased tolerance of *Nicotiana tabacum* to *Tobacco mosaic virus* infection (Masuta et al. 1995). Additionally, pretreatment of *A. thaliana* with CK increased its resistance against *P. syringae* pv. tomato DC3000 (Choi et al. 2010). The enhanced susceptibility of *A. thaliana* SA induction-deficient (*sid2*) mutants to *P. syringae* pv. tomato DC3000 is overcome by CK treatment. This indicates that CK increases plants' immunity in a SA-dependent manner (Naseem et al. 2012). Drought stress leads to decrease in CK levels in plants (Kudoyarova et al. 2007). Genetic modifications leading to low CK levels enhanced plants' resistance to abiotic stresses (Tran et al. 2007, Kang et al. 2012). For example, CK receptor mutant, *Arabidopsis* histidine kinase1 (*ahk1*), exhibits reduced expression of dehydration-responsive elements binding 2A (*DREB2A*) gene and consequently shows decreased tolerance to drought stress (Tran et al. 2007). CKs are known to interact with ABA, SA, and JA signaling (O'Brien and Benková 2013). Different *Arabidopsis* response regulator (ARR) factors that activate the expression of CK-specific genes have been shown to interact with ABI4 and ABI5 and negatively regulate their expression (Wang et al. 2011). It has been shown that JA antagonizes CK-mediated stimulation of cell division and co-regulates CK-mediated increase in cell expansion in *Cucurbita pepo* (Stoynova-Bakalova et al. 2008).

BRs have both positive and negative effects on plant immunity. BRs have been shown to promote resistance to *X. oryzae* and susceptibility to *Pythium graminicola* (causal agent of root rot) in *O. sativa* (De Vleesschauwer et al. 2012, Nakashita et al. 2003). BRs also interact with ABA, SA, and JA (De Bruyne et al. 2014). The exogenous treatment of drought-stressed *S. lycopersicum* with epibrassinolide (EBR) leads to increased ABA levels and activities of antioxidant enzymes like catalase, SOD, and APX, which results in enhanced drought tolerance (Yuan et al. 2010). Whereas BRs are responsible for repressing SA-mediated defense against root pathogens, the hormone enhances the resistance to *X. oryzae* in an SA-independent manner (De Vleesschauwer et al. 2012; Nakashita et al. 2003).

Unlike BRs, exogenous gibberellins are known to enhance susceptibility against *X. oryzae* and resistance against *P. graminicola* in rice (De Vleesschauwer et al. 2012). GAs are known to promote plant growth by inducing the degradation of DELLA proteins. *A. thaliana* quadruple mutants lacking four of the five DELLA proteins showed reduced induction of JA marker gene *PDF1.2*, resulting in increased susceptibility of the plants to necrotrophic fungus *A. brassicicola* (Navarro et al. 2008). GA decreases drought stress tolerance of plants as indicated by a study wherein different mutants with altered GA levels exhibited different sensitivity toward drought. *A. thaliana* gibberellin 20 oxidase 1 (AtGA20ox1) overexpressing plants that accumulate GA is more susceptible to drought stress, whereas knocking out this gene results in improved tolerance to drought stress (Colebrook et al. 2014).

It is evident from the above discussion that response of plants to pathogen is not governed only by SA and JA but is regulated by a complex interplay of all the hormones. The complex interaction among various hormones indicates the tight regulation of plants' defense response against both drought and pathogen infection which suggests further more complexity in regulation in cases where plants are simultaneously challenged with drought and pathogen infection. This suggests the need to study phytohormonal signaling under combined drought and pathogen stress in order to understand plant responses to the combined stress conditions.

9.3 Role of Hormones in Combined Drought and Pathogen Stress Response

The role of hormones in combined stress studies can be understood either directly by imposition of combined drought and pathogen infection on plants or indirectly by using pathogen effectors and exogenous applications of ABA. Both the scenarios are discussed in detail in the following sections.

9.3.1 Clues from Combined Stress Studies

Till date, three transcriptomic studies have been done to understand the molecular response of plants to combined drought and pathogen infection (Atkinson et al. 2013; Prasch and Sonnewald 2013; Choi et al. 2013). These studies indicate the involvement of hormonal signaling in response to combined stress in plants. For example, the transcriptome analysis of *Vitis vinifera* plants exposed to combined drought and *Xylella fastidiosa* (causal agent of Pierce's disease) infection revealed induction of 9-cis-epoxycarotenoid dioxygenase (NCED), two allene oxide synthase paralogs, and three lipoxygenase paralogs, involved in the activation of JA synthesis. In addition, *GO17-U10* gene, which is homologous to two *A. thaliana* genes encoding benzoic acid/SA methyl transferase 1 (*BSMT1*) and JA methyl transferase 1 (*JMT1*) and contributing to the production of methyl salicylate (MeSA) and methyl jasmonate (MeJA), is induced (Choi et al. 2013). The reanalysis of genes specifically induced under combined drought and *Turnip mosaic virus* infection in *A. thaliana* plants (Prasch and Sonnewald 2013) revealed a total of eight genes to be involved in the metabolism of different hormones (Fig. 9.1). Out of these, OPC-8:0 coA ligase1 (*OPCL1*), oxophytodienoate-reductase 3 (*OPR3*), and hydroperoxide lyase 1 (*HPL1*) are involved in JA biosynthesis. Jasmonate-zim-domain protein 2 (*JAZ2*) regulates JA biosynthetic process, small auxin-upregulated 72 (*SAUR72*) and indole-3-acetic acid-inducible 29 (*IAA I29*) are involved in auxin-activated signaling pathway, abscisic acid1 (ABA1) is involved in ABA biosynthetic process, and 4-coumarate:CoA ligase 5 (*4CL5*) is involved in SA biosynthesis. We conclude that tolerance to combined drought and pathogen infection in *A. thaliana* plants involves interaction among at least four hormone pathways.

9.3.2 Clues from Studies Using Pathogen Effectors and Exogenous Application of ABA

The different response of plants to various pathogens is due to interaction of ABA signaling with SA, JA, and ET (Anderson et al. 2004; Xu et al. 2013). Studies with pathogen effectors have shown that pathogens enhance their virulence by modulating ABA signaling. For example, overexpression of *P. syringae* type III effector AvrPtoB led to higher accumulation of ABA in *A. thaliana* (de Torres et al. 2006) leading to enhanced susceptibility of the plant to the pathogen. Overexpression of another type III effector HopAM1, produced by *P. syringae*, led to enhanced pathogen virulence on drought-stressed plants. HopAM1 is also known to modulate the downstream ABA signaling pathways (Goel et al. 2008). Similarly, overexpression of AvrB_AvrC domain of the $AvrXccC_{8004}$ effector from *X. campestris* pv. campestris in *A. thaliana* induced ABA accumulation by upregulating the expression of *NCED5* (Ho et al. 2013). Similarly studies dealing with exogenous application of ABA on pathogen-infected plants have shown that plant-pathogen interaction under drought stress conditions largely depends on the nature of the pathogen that decides the activation of either SA- or JA-mediated responses. The use of pathogen effectors and exogenous ABA treatment can be further useful in scenarios where it is difficult to impose concurrent drought and pathogen infection on plants under laboratory conditions.

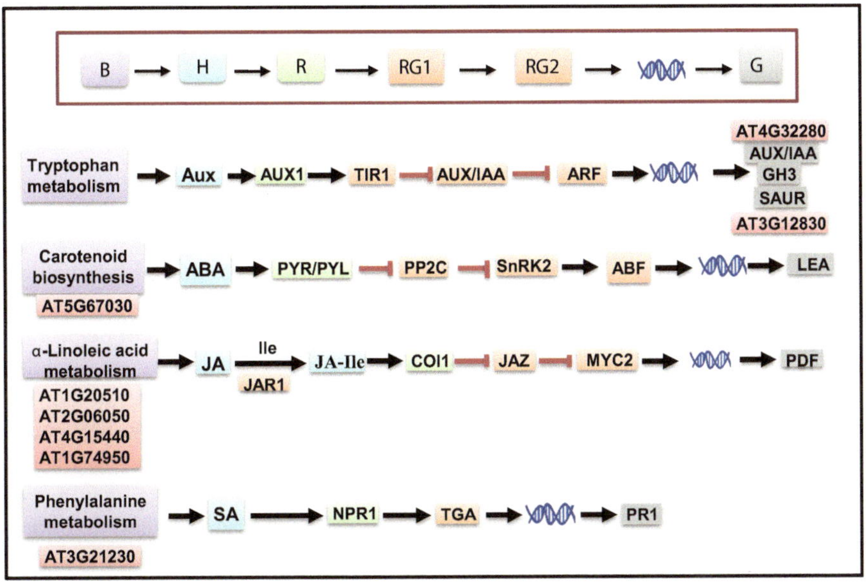

Fig. 9.1 Hypothetical scheme of phytohormonal signaling under combined drought and pathogen infection. The figure represents the signaling pathways of auxin, ABA, JA, and SA with few representative genes starting from biosynthesis of hormone (H) (indicated by B enclosed in *purple box*) their perception by receptors (indicated by R enclosed in *green box*), followed by downstream signaling cascades consisting of regulatory proteins and transcription factors (indicated by RG enclosed in *orange box*), and the gene regulated (indicated by G enclosed in *grey box*). The pathways for auxin, ABA, JA, and SA are shown below in the same order. The TAIR accession IDs correspond to the genes which are specifically upregulated under combined drought and viral infection identified through microarray analysis (Prasch and Sonnewald 2013). *Red blunt* arrows represent inhibition and *black arrows* represent activation. Genes, specifically regulated under combined drought and viral infection, were mapped onto the phytohormone signal transduction pathway database of KEGG. *Aux* auxin, *AUX1* auxin 1, *TIR1* transport inhibitor response1, *AUX/IAA* auxin/indole acetic acid, *ARF* auxin-responsive factor. *GH3* glycoside hydrolase 3, *SAUR* small auxin-up RNAs, *ABA* abscisic acid, *PYR/PYL* pyrabactin resistance, *PP2C* protein phosphatase 2C, *SnRK2* serine/threonine kinase 2, *ABF-ABRE* ABA-responsive element binding factor, *Lea*, late embryogenesis, *JA* jasmonic acid, *Ile* isoleucine, *COI1* coronatine-insensitive 1, *JAZ* jasmonate-zim domain, *MYC2* myelocytomatosis 2. *PDF* plant defensin, *SA* salicylic acid, *NPR1* non-expressor of PR genes1, *PR1* pathogen-responsive 1

9.4 Conclusions and Future Perspectives

The role of ABA, SA, and JA has been established in plant response to individual drought and pathogen infection, and the potential role of these hormones in combined stress response has been indicated recently (Prasch and Sonnewald 2013). Depending on the timing, severity, and lifestyle of the pathogen, a particular hormone signaling pathway might play a major role. However, it is independent hormones' role per se as well as the complex interaction among their signaling with other hormone pathways that tailors plant response to combined stress. Identification of the components of hormonal cross talks as well as the interaction between the signaling pathways can reveal mechanisms behind tolerance to combined drought and pathogen infection. MYC2 transcription factors modulate ABA-SA and ABA-JA cross talk, and NPR1 and WRKY transcription factors regulate the SA-JA cross talk. Studies aiming at characterization of other such modulators can further help in dissecting the complex phytohormonal signaling under combined drought and pathogen infection. Since ABA is a major hormone directly linked to drought stress, studies wherein exogenous application of ABA is followed by challenge with pathogen or pathogen effectors can be used to understand the mechanism of ABA-mediated combined drought and pathogen stress tolerance in plants. This study will also be useful to reveal the complex nature of both drought and pathogen infections, where it is difficult at times to coincide drought stress treatment with pathogen infection under laboratory conditions. Identification of molecular players important in conferring combined drought and pathogen stress tolerance to plants by transcriptomic studies and functional analysis of mutants for the genes identified can also help in unraveling the associated signaling pathways. Altogether, uncovering the complex phytohormonal signaling under the combined stress conditions will not only help in enriching our understanding of plants' defense responses to the stress conditions but also lead to identifying potential candidate genes for improving combined drought and pathogen stress tolerance of plants.

Acknowledgments Combined stress tolerance-related projects at MS-K Lab are supported by the National Institute of Plant Genome Research core funding and the DBT-Ramalingaswami Re-entry Fellowship grant (BT/RLF/re-entry/23/2012) and the DBT-Innovative Young Biotechnologist award. PP acknowledges DST-SERB Young Scientist grant (SB/YS/LS-71/2014) for the financial support.

References

Abe H, Urao T, Ito T, Seki M, Shinozaki K, Yamaguchi-Shinozaki K (2003) Arabidopsis AtMYC2 (bHLH) and AtMYB2 (MYB) function as transcriptional activators in abscisic acid signaling. Plant Cell 15:63–78

Adie BAT, Pérez-Pérez J, Pérez-Pérez MM, Godoy M, Sánchez-Serrano J-J, Schmelz EA, Solano R (2007) ABA is an essential signal for plant resistance to pathogens affecting JA biosynthesis and the activation of defenses in Arabidopsis. Plant Cell 19:665–681

Alam M, Nahar K, Hasanuzzaman M, Fujita M (2014) Exogenous jasmonic acid modulates the physiology, antioxidant defense and glyoxalase systems in imparting drought stress tolerance in different Brassica species. Plant Biotechnol Rep 8:279–293

Anderson JP, Badruzsaufari E, Schenk PM, Manners JM, Desmond OJ, Ehlert C (2004) Antagonistic interaction between abscisic acid and jasmonate-ethylene signaling pathways modulates defense gene expression and disease resistance in Arabidopsis. Plant Cell 16:3460–3479

Asselbergh B, De Vleesschauwer D, Höfte M (2008) Global switches and fine-tuning-ABA modulates plant-pathogen defense. Mol Plant-Microbe Interact 21:709–719

Atkinson NJ, Lilley CJ, Urwin PE (2013) Identification of genes involved in the response of Arabidopsis to simultaneous biotic and abiotic stresses. Plant Physiol 162:2028–2041

Audenaert K, De Meyer GB, Höfte MM (2002) Abscisic acid determines basal susceptibility of tomato to *Botrytis cinerea* and suppresses salicylic acid-dependent signaling mechanisms. Plant Physiol 128:491–501

Bandurska H, Stroiński A (2005) The effect of salicylic acid on barley response to water deficit. Acta Physiol Plant 27:379–386

Borsani O, Valpuesta V, Botella MA (2001) Evidence for a role of salicylic acid in the oxidative damage generated by NaCl and osmotic stress in Arabidopsis seedlings. Plant Physiol 126:1024–1030

Bowling SA, Clarke JD, Liu Y, Klessig DF, Dong X (1997) The cpr5 mutant of Arabidopsis expresses both NPR1-dependent and NPR1-independent resistance. Plant Cell 9:1573–1584

Chen Z, Agnew JL, Cohen JD, He P, Shan L, Sheen J, Kunkel BN (2007) *Pseudomonas syringae* type III effector AvrRpt2 alters *Arabidopsis thaliana* auxin physiology. Proc Natl Acad Sci USA 104:20131–20136

Chen H, Lai Z, Shi J, Xiao Y, Chen Z, Xu X (2010) Roles of Arabidopsis WRKY18, WRKY40 and WRKY60 transcription factors in plant responses to abscisic acid and abiotic stress. BMC Plant Biol 10:281

Chini A, Grant JJ, Seki M, Shinozaki K, Loake GJ (2004) Drought tolerance established by enhanced expression of the CC-NBS-LRR gene, ADR1, requires salicylic acid, EDS1 and ABI1. Plant J 38(5):810–822

Choi J, Huh SU, Kojima M, Sakakibara H, Paek K-H, Hwang I (2010) The cytokinin-activated transcription factor ARR2 promotes plant immunity via TGA3/NPR1-dependent salicylic acid signaling in Arabidopsis. Dev Cell 19:284–295

Choi HK, Iandolino A, da Silva FG, Cook DR (2013) Water deficit modulates the response of *Vitis vinifera* to the Pierce's disease pathogen *Xylella fastidiosa*. Mol Plant-Microbe Interact 26:643–657

Colebrook EH, Thomas SG, Phillips AL, Hedden P (2014) The role of gibberellin signalling in plant responses to abiotic stress. J Exp Biol 217:67–75

De Bruyne L, Höfte M, De Vleesschauwer D (2014) Connecting growth and defense: the emerging roles of brassinosteroids and gibberellins in plant innate immunity. Mol Plant 7(6):943–959

De Ollas C, Hernando B, Arbona V, Gómez-Cadenas A (2013) Jasmonic acid transient accumulation is needed for abscisic acid increase in citrus roots under drought stress conditions. Physiol Plant 147:296–306

De Ollas C, Arbona V, Gómez-Cadenas A (2015) Jasmonoyl isoleucine accumulation is needed for abscisic acid build-up in roots of Arabidopsis under water stress conditions. Plant Cell Environ 38:2157–2170

de Torres M, Mansfield JW, Grabov N, Brown IR, Ammouneh H, Tsiamis G, Forsyth A, Robatzek S, Grant M, Boch J (2006) *Pseudomonas syringae* effector AvrPtoB suppresses basal defence in Arabidopsis. Plant J 47:368–382

de Torres-Zabala M, Bennett MH, Truman WH, Grant MR (2009) Antagonism between salicylic and abscisic acid reflects early host–pathogen conflict and moulds plant defence responses. Plant J 59:375–386

de Torres-Zabala M, Truman W, Bennett MH, Lafforgue G, Mansfield JW, Rodriguez Egea P, Bögre L, Grant M (2007) Pseudomonas syringae pv. tomato hijacks the Arabidopsis abscisic acid signalling pathway to cause disease. EMBO J 26(5):1434–1443

De Vleesschauwer D, Yang YN, Cruz CV, Höfte M (2010) Abscisic acid-induced resistance against the brown spot pathogen *Cochliobolus miyabeanus* in rice involves MAP kinase-mediated repression of ethylene signaling. Plant Physiol 152:2036–2052

De Vleesschauwer D, Van Buyten E, Satoh K, Balidion J, Mauleon R, Choi IR, Vera-Cruz C, Kikuchi S, Höfte M (2012) Brassinosteroids antagonize gibberellin- and salicylate-mediated root immunity in rice. Plant Physiol 158:1833–1846

De Vleesschauwer D, Xu J, Höfte M (2014) Making sense of hormone-mediated defense networking: from rice to Arabidopsis. Front Plant Sci 5:611

Du H, Liu H, Xiong L (2013) Endogenous auxin and jasmonic acid levels are differentially modulated by abiotic stresses in rice. Front Plant Sci 4:397

Fan J, Hill L, Crooks C, Doerner P, Lamb C (2009) Abscisic acid has a key role in modulating diverse plant-pathogen interactions. Plant Physiol 150(4):1750–1761

Feng DX, Tasset C, Hanemian M, Barlet X, Hu J, Trémousaygue D, Deslandes L, Marco Y (2012) Biological control of bacterial wilt in Arabidopsis thaliana involves abscisic acid signalling. New Phytol 194(4):1035–1045

Fernández-Calvo P, Chini A, Fernández-Barbero G, Chico JM, Gimenez-Ibanez S, Geerinck J, Eeckhout D, Schweizer F, Godoy M, Franco-Zorrilla JM, Pauwels L, Witters E, Puga MI, Paz-Ares J, Goossens A, Reymond P, De Jaeger G, Solano R (2011) The Arabidopsis bHLH transcription factors MYC3 and MYC4 are targets of JAZ repressors and act additively with MYC2 in the activation of jasmonate responses. Plant Cell 23:701–715

Fu J, Wang S (2011) Insights into auxin signaling in plant-pathogen interactions. Front Plant Sci 2:1–7

Fu J, Liu H, Li Y, Yu H, Li X, Xiao J, Wang S (2011) Manipulating broad-spectrum disease resistance by suppressing pathogen-induced auxin accumulation in rice. Plant Physiol 155:589–602

Fujita M, Fujita Y, Noutoshi Y, Takahashi F, Narusaka Y, Yamaguchi-Shinozaki K, Shinozaki K (2006) Crosstalk between abiotic and biotic stress responses: a current view from the points of convergence in the stress signaling networks. Curr Opin Plant Biol 9:436–442

Gao XP, Pan QH, Li MJ, Zhang LY, Wang XF, Shen YY, Lu YF, Chen SW, Liang Z, Zhang DP (2004) Abscisic acid is involved in the water stress-induced betaine accumulation in pear leaves. Plant Cell Physiol 45(6):742–750

Gao QM, Venugopal S, Navarre D, Kachroo A (2011) Low oleic acid-derived repression of jasmonic acid-inducible defense responses requires the WRKY50 and WRKY51 proteins. Plant Physiol 155:464–476

Glazebrook J (2005) Contrasting mechanisms of defense against biotrophic and necrotrophic pathogens. Annu Rev Phytopathol 43:205–227

Goel AK, Lundberg D, Torres MA, Matthews R, Akimoto-Tomiyama C, Farmer L, Dangl JL, Grant SR (2008) The *Pseudomonas syringae* type III effector HopAM1 enhances virulence on water stressed plants. Mol Plant-Microbe Interact 21:361–370

Grant JJ, Chini A, Basu D, Loake GJ (2003) Targeted activation tagging of the Arabidopsis NBS-LRR gene, ADR1, conveys resistance to virulent pathogens. Mol Plant-Microbe Interact 16:669–680

Harb A, Krishnan A, Ambavaram MMR, Pereira A (2010) Molecular and physiological analysis of drought stress in Arabidopsis reveals early responses leading to acclimation in plant growth. Plant Physiol 154:1254–1271

Hernández-Blanco C, Feng DX, Hu J, Sánchez-Vallet A, Deslandes L, Llorente F, Berrocal-Lobo M, Keller H, Barlet X, Sánchez-Rodríguez C, Anderson LK, Somerville S, Marco Y, Molina A (2007) Impairment of cellulose synthases required for Arabidopsis secondary cell wall formation enhances disease resistance. Plant Cell 19(3):890–903

Higashi K, Ishiga Y, Inagaki Y, Toyoda K, Shiraishi T, Ichinose Y (2008) Modulation of defense signal transduction by flagellin-induced WRKY41 transcription factor in Arabidopsis thaliana. Mol Gen Genomics 279:303–312

Ho YP, Tan CM, Li MY, Lin H, Deng WL, Yang JY (2013) The AvrB_AvrC domain of AvrXccC of *Xanthomonas campestris* pv. campestris is required to elicit plant defense responses and manipulate ABA homeostasis. Mol Plant Microbe Interact 26(4):419–430

Jiang CJ, Shimono M, Sugano S, Kojima M, Yazawa K, Yoshida R, Inoue H, Hayashi N, Sakakibara H, Takatsuji H (2010) Abscisic acid interacts antagonistically with salicylic acid signaling pathway in rice-*Magnaporthe grisea* interaction. Mol Plant-Microbe Interact 23(6):791–798

Journot-Catalino N, Somssich IE, Roby D, Kroj T (2006) The transcription factors WRKY11 and WRKY17 act as negative regulators of basal resistance in *Arabidopsis thaliana*. Plant Cell 18:3289–3302

Kang NY, Cho C, Kim NY, Kim J (2012) Cytokinin receptor-dependent and receptor-independent pathways in the dehydration response of *Arabidopsis thaliana*. J Plant Physiol 169:1382–1391

Kazan K (2015) Diverse roles of jasmonates and ethylene in abiotic stress tolerance. Trends Plant Sci 20(4):219–229

Koga H, Dohia K, Moria M (2004) Abscisic acid and low temperatures suppress the whole plant-specific resistance reaction of rice plants to the infection of *Magnaporthe grisea*. Physiol Mol Plant Pathol 65:3–9

Korkmaz A, Uzunlu M, Demirkiran A (2007) Treatment with acetyl salicylic acid protects muskmelon seedlings against drought stress. Acta Physiol Plant 29:503–508

Kudoyarova GR, Vysotskaya LB, Cherkozyanova A, Dodd IC (2007) Effect of partial root zone drying on the concentration of zeatin-type cytokinins in tomato (*Solanum lycopersicum* L.) xylem sap and leaves. J Exp Bot 58:161–168

Leon-Reyes A, Van der Does D, De Lange ES, Delker C, Wasternack C, Van Wees SC, Ritsema T, Pieterse CM (2010) Salicylate-mediated suppression of jasmonate-responsive gene expression in Arabidopsis is targeted downstream of the jasmonate biosynthesis. Planta 232(6):1423–1432

Li J, Brader G, Kariola T, Palva ET (2006) WRKY70 modulates the selection of signaling pathways in plant defense. Plant J 46(3):477–491

Lorenzo O, Piqueras R, Sánchez-Serrano JJ, Solano R (2003) ETHYLENE RESPONSE FACTOR1 integrates signals from ethylene and jasmonate pathways in plant defense. Plant Cell 15:165–178

Lorenzo O, Chico JM, Sánchez-Serrano JJ, Solano R (2004) JASMONATE-INSENSITIVE1 encodes a MYC transcription factor essential to discriminate between different jasmonate-regulated defense responses in Arabidopsis. Plant Cell 16:1938–1950

Mao P, Duan M, Wei C, Li Y (2007) WRKY62 transcription factor acts downstream of cytosolic NPR1 and negatively regulates jasmonate-responsive gene expression. Plant Cell Physiol 48:833–842

Masuta C, Tanaka H, Uehara K, Kuwata S, Koiwai A, Noma M (1995) Broad resistance to plant viruses in transgenic plants conferred by antisense inhibition of a host gene essential in S-adenosylmethionine-dependent transmethylation reactions. Proc Natl Acad Sci USA 92:6117–6121

Mauch-Mani B, Mauch F (2005) The role of abscisic acid in plant-pathogen interactions. Curr Opin Plant Biol 8:409–414

Melotto M, Underwood W, Koczan J, Nomura K, He SY (2006) Plant stomata function in innate immunity against bacterial invasion. Cell 126:969–980

Mengiste T, Chen X, Salmeron J, Dietrich R (2003) The BOTRYTIS SUSCEPTABLE1 gene encodes an R2R3MYB transcription factor protein that is required for biotic and abiotic stress responses in Arabidopsis. Plant Cell 15:2551–2565

Miura K, Tada Y (2014) Regulation of water, salinity, and cold stress responses by salicylic acid. Front Plant Sci 5:4

Miura K, Okamoto H, Okuma E, Shiba H, Kamada H, Hasegawa PM, Murata Y (2013) SIZ1 deficiency causes reduced stomatal aperture and enhanced drought tolerance via controlling salicylic acid-induced accumulation of reactive oxygen species in Arabidopsis. Plant J 49:79–90

Mohr PG, Cahill DM (2003) Abscisic acid influences the susceptibility of *Arabidopsis thaliana* to *Pseudomonas syringae* pv. tomato and *Peronospora parasitica*. Funct Plant Biol 30:461–469

Mohr PG, Cahill DM (2007) Suppression by ABA of salicylic acid and lignin accumulation and the expression of multiple genes, in Arabidopsis infected with *Pseudomonas syringae* pv. tomato. Funct Integr Genomics 7(3):181–191

Munemasa S, Oda K, Watanabe-Sugimoto M, Nakamura Y, Shimoishi Y, Murata Y (2007) The coronatine-insensitive 1 mutation reveals the hormonal signaling interaction between abscisic acid and methyl jasmonate in Arabidopsis guard cells. Specific impairment of ion channel activation and second messenger production. Plant Physiol 143:1398–1407

Munne-Bosch S, Penuelas J (2003) Photo- and antioxidative protection, and a role for salicylic acid during drought and recovery in field-grown *Phillyrea angustifolia* plants. Planta 217:758–766

Mur LAJ, Kenton P, Atzorn R, Miersch O, Wasternack C (2006) The outcomes of concentration-specific interactions between salicylate and jasmonate signaling include synergy, antagonism, and oxidative stress leading to cell death. Plant Physiol 140:249–262

Nakashita H, Yasuda M, Nitta T, Asami T, Fujioka S, Arai Y, Sekimata K, Takatsuto S, Yamaguchi I, Yoshida S (2003) Brassinosteroid functions in a broad range of disease resistance in tobacco and rice. Plant J 33:887–889

Naseem M, Philippi N, Hussain A, Wangorsch G, Ahmed N, Dandekar T (2012) Integrated systems view on networking by hormones in Arabidopsis immunity reveals multiple crosstalk for cytokinin. Plant Cell 24:1793–1814

Navarro L, Bari R, Achard P, Lisón P, Nemri A, Harberd NP, Jones JDG (2008) DELLAs control plant immune responses by modulating the balance of jasmonic acid and salicylic acid signaling. Curr Biol 18:650–655

O'Brien JA, Benková E (2013) Cytokinin cross-talking during biotic and abiotic stress responses. Front Plant Sci 4:451

Okuma E, Nozawa R, Murata Y, Miura K (2014) Accumulation of endogenous salicylic acid confers drought tolerance to Arabidopsis. Plant Signal Behav 9(3):e28085

Pandey P, Ramegowda V, Senthil-Kumar M (2015) Shared and unique responses of plants to multiple individual stresses and stress combinations: physiological and molecular mechanisms. Front Plant Sci 6:723

Peña-Cortés H, Fisahn J, Willmitzer L (1995) Signals involved in wound-induced proteinase inhibitor II gene expression in tomato and potato plants. Proc Natl Acad Sci USA 92:4106–4113

Pieterse CMJ, Van der Does D, Zamioudis C, Leon-Reyes A, Van Wees SCM (2012) Hormonal modulation of plant immunity. Annu Rev Cell Dev Biol 28:489–521

Prasch CM, Sonnewald U (2013) Simultaneous application of heat, drought and virus to Arabidopsis thaliana plants reveals significant shifts in signaling networks. Plant Physiol 162(4):1849–1866

Qiu Y, Yu D (2009) Over-expression of the stress-induced OsWRKY45 enhances disease resistance and drought tolerance in Arabidopsis. Environ Exp Bot 65:35–44

Ramegowda V, Senthil-kumar M (2015) The interactive effects of simultaneous biotic and abiotic stresses on plants: mechanistic understanding from drought and pathogen combination. J Plant Physiol 176:47–54

Rasmussen S, Barah P, Suarez-Rodriguez MC, Bressendorff S, Friis P, Costantino P, Bones AM, Nielsen HB, Mundy J (2013) Transcriptome responses to combinations of stresses in Arabidopsis. Plant Physiol 161:1783–1794

Sanchez-Vallet A, López G, Ramos B, Delgado-Cerezo M, Riviere MP, Llorente F, Fernández PV, Miedes E, Estevez JM, Grant M, Molina A (2012) Disruption of abscisic acid signaling constitutively activates Arabidopsis resistance to the necrotrophic fungus *Plectosphaerella cucumerina*. Plant Physiol 160(4):2109–2124

Saruhan N, Saglam A, Kadioglu A (2012) Salicylic acid pretreatment induces drought tolerance and delays leaf rolling by inducing antioxidant systems in maize genotypes. Acta Physiol Plant 34:97–106

Spoel SH, Dong X (2008) Making sense of hormone crosstalk during plant immune responses. Cell Host Microbe 3(6):348–351

Spoel SH, Koornneef A, Claessens SMC, Korzelius JP, Van Pelt JA, Mueller MJ, Buchala AJ, Métraux JP, Brown R, Kazan K, Van Loon LC, Dong X, Pieterse CM (2003) NPR1 modulates cross-talk between salicylate- and jasmonate-dependent defense pathways through a novel function in the cytosol. Plant Cell 15:760–770

Spoel SH, Johnson JS, Dong X (2007) Regulation of tradeoffs between plant defenses against pathogens with different lifestyles. Proc Natl Acad Sci USA 104:18842–18847

Stoynova-Bakalova E, Petrov PI, Gigova L, Baskin TI (2008) Differential effects of methyl jasmonate on growth and division of etiolated zucchini cotyledons. Plant Biol 10:476–484

Suzuki N, Rivero RM, Shulaev V, Blumwald E, Mittler R (2014) Abiotic and biotic stress combinations. New Phytol 203(1):32–43

Ton J, Flors V, Mauch-Mani B (2009) The multifaceted role of ABA in disease resistance. Trends Plant Sci 14:10–317

Tran L-SP, Urao T, Qin F, Maruyama K, Kakimoto T, Shinozaki K, Yamaguchi-Shinozaki K (2007) Functional analysis of AHK1/ATHK1 and cytokinin receptor histidine kinases in response to abscisic acid, drought, and salt stress in Arabidopsis. Proc Natl Acad Sci USA 104:20623–20628

Ueno Y, Yoshida R, Kishi-Kaboshi M, Matsushita A, Jiang C-J, Goto S, Takahashi A, Hirochika H, Takatsuji H (2015) Abiotic stresses Antagonize the rice defence pathway through the Tyrosine-Dephosphorylation of OsMPK6. PLoS Pathog 11(10):e1005231

Ulferts S, Delventhal R, Splivallo R, Karlovsky P, Schaffrath U (2015) Abscisic acid negatively interferes with basal defence of barley against *Magnaporthe oryzae*. BMC Plant Biol 15:7

Wang Y, Li L, Ye T, Zhao S, Liu Z, Feng YQ, Wu Y (2011) Cytokinin antagonizes ABA suppression to seed germination of Arabidopsis by downregulating ABI5 expression. Plant J 68:249–261

Wiese J, Kranz T, Schubert S (2004) Induction of pathogen resistance in barley by abiotic stress. Plant Biol (Stuttg) 6(5):529–536

Xu X, Chen C, Fan B, Chen Z (2006) Physical and functional interactions between pathogen-induced Arabidopsis WRKY18, WRKY40, and WRKY60 transcription factors. Plant Cell 18:1310–1326

Xu J, Audenaert K, Höfte M, De Vleesschauwer D (2013) Abscisic acid promotes susceptibility to the rice leaf blight pathogen *Xanthomonas oryzae* pv oryzae by suppressing salicylic acid-mediated defenses. PLoS One 8:e67413

Yuan G-F, Jia C-G, Li Z, Sun B, Zhang L-P, Liu N, Wang Q-M (2010) Effect of brassinosteroids on drought resistance and abscisic acid concentration in tomato under water stress. Sci Hort 126:103–108

Zhang SW, Li CH, Cao J, Zhang YC, Zhang SQ, Xia YF, Sun DY, Sun Y (2009) Altered architecture and enhanced drought tolerance in rice via the down-regulation of indole-3-acetic acid by TLD1/OsGH3.13 activation. Plant Physiol 151(4):1889–1901

Zheng XY, Spivey NW, Zeng W, Liu PP, Fu ZQ, Klessig DF, He SY, Dong X (2012) Coronatine promotes *Pseudomonas syringae* virulence in plants by activating a signaling cascade that inhibits salicylic acid accumulation. Cell Host Microbe 11(6):587–596

Zhu Z, Fengying A, Ying F, Pengpeng L, Li X, Mu A, Zhiqiang J, Jong-Myong K, Taiko KT, Wei L, Xinyan Z, Qiang Y, Zhi D, Wen-Qian C, Motoaki S, Jian-Min Z, Hongwei G (2011) Derepression of ethylene-stabilized transcription factors (EIN3/EIL1) mediates jasmonate and ethylene signaling synergy in Arabidopsis. Proc Natl Acad Sci USA 108:12539–12544

Simultaneous Expression of Abiotic Stress-Responsive Genes: An Approach to Improve Multiple Stress Tolerance in Crops

10

M. S. Parvathi and Karaba N. Nataraja

Abstract

Survival and productivity of crop plants under different stress scenarios are critical for sustainable crop production. Crop improvement towards achieving higher yield potentials under multiple stresses requires targeted breeding of crop varieties that can withstand different levels of abiotic and biotic stresses and still produce a reliable yield. In addition to this, many candidate genes have been validated for studying their role in enhancing tolerance levels in crops. Transgenic development with targeted stress tolerance trait manipulation can be a viable approach for attaining multiple stress tolerance. By this method, multilevel stress tolerance can be achieved by either manipulating candidate regulatory genes which can govern the functionality of a range of downstream genes or by adopting multigene stacking approach to pyramid genes in a same genetic background. The careful selection of genes linked to specific traits is the key to achieve marginal increase in productivity under multiple stresses. Diverse genes associated with various cellular tolerance mechanisms act in a concerted manner to impart varying degrees of stress tolerance. It will be highly rewarding if we examine different pathway-linked genes active in the stress scenario under scrutiny. Targeted genetic manipulation to enhance cellular tolerance under stress will be more economically viable if we combine multiple trait regulatory genes by using modern biotechnological tools. The different strategies employed, advantages of simultaneous expression of trait regulatory genes and the resultant crop adaptive mechanisms that are emerging in the recent years are discussed in this chapter.

M.S. Parvathi • K.N. Nataraja (✉)
Plant Molecular Biology Laboratory, Department of Crop Physiology, University of Agricultural Sciences, Gandhi Krishi Vigyan Kendra, Bengaluru 560065, India
e-mail: nataraja_karaba@yahoo.com; nnkaraba@uasbangalore.edu.in

M. Senthil-Kumar (eds.), *Plant Tolerance to Individual and Concurrent Stresses*,
DOI 10.1007/978-81-322-3706-8_10

Keywords
Abiotic stress • Co-expression • Traits • Multiple stress tolerance

10.1 Introduction

Global climate change caused by anthropogenic activities is a threat to the sustained food production in developing countries. Food security can be achieved by adopting efficient measures to mitigate the negative effects of environmental stresses. Multiple abiotic stresses adversely affect crop productivity in tropical regions. Under such circumstances, it is crucial to ensure that food requirements are met through enhanced productivity with sustained nutritional security. Crop improvement towards achieving realisable levels of stress tolerance requires a holistic strategy by integrating multiple stress tolerance mechanisms. This can be addressed by identifying and analysing the decisive traits that can be selected for trait manipulation by new plant breeding molecular technologies and other biotechnological approaches.

10.2 Common Stresses Encountered in Tropical Regions: Drought, Heat and High Light

Abiotic stresses in combination cause both general and specific detrimental effects on plant growth and productivity. In tropical conditions, in most cases, drought, high temperature and high light stresses occur simultaneously. Crop productivity is hindered under natural conditions where combinations of two or more stresses, such as drought and salinity and salinity and heat, and combinations of drought with extreme temperature or high light intensity exist. A recent report on the comparison of all major US weather disasters between 1980 and 2012 indicates that a combination of drought and heat stress caused extensive agricultural losses worth approximately $200 billion. In the same period, drought alone caused $50 billion worth of damage to agricultural production (Mittler 2006; Suzuki et al. 2014).

There are various tolerance/acclimation mechanisms which integrate to alter gene expression and/or protein functionalities, thereby paving way for either avoidance or tolerance strategies for plant survival. Stress tolerance is also governed by duration, rate, severity and combinatorial effects of stress conditions along with genotypic as well as plant developmental stage variations. There are certain complex physiological traits which are vital for multiple stress tolerance (Kamoshita et al. 2008). The mechanism of adaptation and morphological responses to drought, heat and high light varies considerably. For example, water mining and water conservation strategies are common under drought (Wasson et al. 2012). In many crop plants, drought-induced root growth has been reported as an efficient water mining strategy (Comas et al. 2013). Water conservation is achieved by reduced leaf number and area and also increased stomatal and cuticular resistances (Mamrutha et al. 2010). On the other hand, plants also have long-term adaptive strategies for heat tolerance, mainly by altered metabolic process and decreased leaf canopy temperature through increased transpiration (Wahid et al. 2007). Heat and drought may have contrasting effects on a few morphological features. For example, drought-adaptive strategy involves reduced leaf number and leaf expansion/area (Alves and Setter 2004; Poorter et al. 2009), while in heat acclimation, increase in leaf number and leaf elongation has been reported (Bos et al. 2000; Wahid et al. 2007). Heat stress might reduce the number, length and diameter of roots, and moderate drought stress increases root growth, a water mining trait (Wahid et al. 2007). There are reports on increased biomass allocations to roots under drought, while heat stress might enhance reproductive allocation. There is complex interaction under field conditions, and hence some of the responses are believed to be shared among drought and high temperature stresses. In *Arabidopsis thaliana*, during combined drought and high temperature stress, leaf size was found to increase with decreased leaf number, and biomass allocation was more towards roots and reproductive parts (Vile et al. 2012).

When light energy exceeds the photosynthetic capacity, it induces photoinhibition, which is common under drought and high temperature conditions. Under drought, the effect is pronounced due to limitation of substrate CO_2 caused by partial closure of stomata. Photoinhibition occurs mainly in photosystem II (PSII), although photoinhibition of photosystem I (PSI) has also been reported in some species (Essemine et al. 2012). Over reduction of the electron transport chain results in the generation of reactive oxygen species (ROS, superoxide anion radicals or/and singlet oxygen) leading to oxidative stress. Plants acclimate to high light stress by dissipating the excess energy as heat and non-photochemical quenching of chlorophyll fluorescence, NPQ (Niyogi et al. 2005; Demmig-Adams and Adams 2006). Understanding the mechanisms or the traits associated with prevention and detoxification of ROS is essential to minimise the ill effects of oxidative stress during combined stresses.

Among the key traits contributing for combined stress tolerance, cellular tolerance (CT) forms a major trait which can impart tolerance to multiple stresses. Many CT trait-specific genes have been identified, which fall into different categories like those involved in stress signal perception, amplification and transduction leading to activation of transcription factors which regulate the expression of downstream functional genes (Parvathi and Nataraja 2016). The common genes expressed under combined drought and heat stress such as those encoding heat shock proteins (HSPs), ROS detoxification enzymes, late embryogenesis abundant 7 (LEA7), dehydrins and genes encoding enzymes involved in pentose pathway and anthocyanin biosynthesis (Rizhsky et al. 2002, 2004; Rampino et al. 2006) contribute for maintaining cellular activities under stress. Rizhsky et al. (2002) reported induction of genes encoding small HSPs (sHSPs), like HSP70, HSP90 and HSP100 under individual as well as combined stress in tobacco. Cellular tolerance mechanisms activated under different abiotic stresses seem to have common regulatory pathways. For example, some of the WRKY transcription factors (TFs) and ethylene response transcriptional co-activator (ERTCA), receptor-like kinases, protein kinases (MAP3K), small GTP-binding proteins and MYB TFs have been reported to be activated under combined stress (Rizhsky et al. 2002, 2004; Rampino et al. 2006). The existing information suggests that there could be a possibility of manipulating diverse CT pathways using the candidate regulatory genes. Therefore, pyramiding or stacking regulatory genes linked to multiple stress tolerance would be useful for trait improvement.

10.3 Evolution of Gene Stacking Approach to Combine Abiotic Stress Tolerance Traits

Crop improvement towards achieving higher yield potentials under different abiotic stresses requires sustained breeding of crop varieties that can withstand different levels of stresses and still produce sufficiently (Tirado and Cotter 2010). The different methods currently being used to develop drought-tolerant crops include combining traits through conventional breeding, conventional breeding utilising marker-assisted selection (MAS) and genetic engineering using trait-specific genes (Tirado and Cotter 2010). In addition, new plant breeding molecular technologies are being evolved and would be useful to manipulate the relevant traits in a precise manner. There is a need to combine traits to achieve the goal of developing crops tolerant to multiple stresses.

Genetic manipulation or engineering of many functionally relevant candidate genes has been undertaken to test their effects on enhancing cellular tolerance levels in crops by over expression and suppression of transgenic technologies as reviewed extensively in the past few years (Lawlor 2013; Hu and Xiong 2014). Transgenic technology has been used extensively to develop crop varieties or cultivars to attain tolerance to pests and diseases, herbicides and biotic and abiotic stresses and for other purposes like nutritional enhancement and phytoremediation (Bakshi and Dewan 2013). Recent reviews have also highlighted the increase in transgenics developed either for attaining higher abiotic stress tolerance potentials in cereals (especially rice) or for functional validation of relevant candidate genes (Hadiarto and Tran 2011; Lawlor 2013; Todaka et al. 2015). Hence, transgenic development with targeted trait manipulation can be a viable approach for attaining multiple abiotic stress tolerance (Jacobsen and Nataraja 2008). The success in these approaches depends on the identification of the trait/s to be manipulated and careful selection of gene/s underlying component tolerance trait/s. Transcription factors are the most commonly employed genes for crop improvement for altering different cellular tolerance traits in the recent past, as enlisted in Table 10.1.

Although single gene transgenics have been successful to improve plant tolerance levels (as reviewed in Reguera et al. 2012 and Lawlor 2013), the alteration of complex interactive metabolic pathways and important quantitative traits requires co-expression of multiple genes (Vemanna et al. 2013). The major bottleneck in manipulating complex traits has been the technical limitations to introduce multiple genes into plants. Gene stacking in general refers to the process of combining two or more trait regulatory genes into a single crop plant. Gene pyramiding and multigene transfer are the synonyms for the same process (ISAAA 2013). Gene stacking

Table 10.1 Different stress-responsive transcription factors used for enhancing abiotic stress tolerance in model and crop plants

Transcription factor family	Gene name	Source	Species manipulated	Stress tolerance[a]	References
bZIP	*GmbZIP1*	*Glycine max*	Arabidopsis	D, S, C	Gao et al. (2011)
	TabZIP60	*Triticum aestivum*	Arabidopsis	D, S, F	Zhang et al. (2015b)
	OsbZIP71	*Oryza sativa*	Rice	D, S	Liu et al. (2014)
	EcbZIP60	*Eleusine coracana*	Tobacco	D, S, O, E	Babitha et al. (2015a)
NAC	*TaNAC29*	*Triticum aestivum*	Arabidopsis	D, S	Huang et al. (2015)
	SNAC1	*Oryza sativa*	Wheat	D, S	Saad et al. (2013)
	OsNAP	*Oryza sativa*	Rice	C, S, D	Chen et al. (2014)
	AhNAC3	*Arachis hypogaea*	Tobacco	D	Liu et al. (2013)
	ZmSNAC1	*Zea mays*	Arabidopsis	C, S, D	Lu et al. (2012)
	EcNAC1	*Eleusine coracana*	Tobacco	Os, S, O	Ramegowda et al. (2012)
	AtNAC2	*Arabidopsis thaliana*	Peanut	D, S	Patil et al. (2014)
WRKY	*ZmWRKY58*	*Zea mays*	Rice	D, S	Cai et al. (2014)
	MtWRKY76	*Medicago truncatula*	Alfalfa	D, S	Liu et al. (2016)
	TaWRKY93	*Triticum aestivum*	Arabidopsis	S, D, LT	Qin et al. (2015)
	OsWRKY45	*Oryza sativa*	Arabidopsis	D, S	Qiu and Yu (2009)
	GhWRKY39	*Gossypium hirsutum*	Tobacco	S	Shi et al. (2014)
MYB	*OsMYB91*	*Oryza sativa*	Rice	S	Zhu et al. (2015)
	LeAN2	*Solanum lycopersicum*	Tomato	H	Meng et al. (2015)
	TaPIMP1	*Triticum aestivum*	Tobacco	D, S	Liu et al. (2011)
	TaMYB3R1	*Triticum aestivum*	Arabidopsis	D, S	Cai et al. (2015)
AP2/ERFBP	*StDREB1*	*Solanum tuberosum*	Potato	S	Bouaziz et al. (2013)
	TaERF3	*Triticum aestivum*	Wheat	D, S	Rong et al. (2014)
	AtDREB1A	*Arabidopsis thaliana*	Rice	D	Ravikumar et al. (2014)
	JERF3	*Solanum lycopersicum*	Rice	D	Zhang et al. (2010)
	OsERF4a	*Oryza sativa*	Rice	D	Joo et al. (2013)
bHLH	*bHLH92*	*Arabidopsis thaliana*	Arabidopsis	S, Os	Jiang et al. (2009)
	AtICE1	*Arabidopsis thaliana*	Tobacco	D, S, Os	Budhagatapalli et al. (2016)
	MdCIbHLH1	*Malus domestica*	Arabidopsis	C	Feng et al. (2012)
	EcbHLH57	*Eleusine coracana*	Tobacco	S, D	Babitha et al. (2015b)

[a]*D* drought, *S* salinity, *O* oxidative stress, *Os* osmotic stress, *C* cold, *F* freezing stress, *H* heat, *LT* low temperature, *E* ER stress

in transgenic plants was addressed as the challenge for twenty-first-century plant biotechnology in 2005 (Halpin 2005). There has been a clear-cut evolution of gene stacking strategies which have been developed over the past one and a half decades. There were a few iterative strategies such as (a) hybrid stacking (crossing two transgenic plants harbouring different transgenes) and (b) retransformation (plant harbouring a transgene is transformed with new transgenes) (Halpin 2005; ISAAA 2013). Later on, taking into consideration the limitations of these iterative strategies, a variety of conventional and novel techniques have been evolved (Halpin 2005; ISAAA 2013). The common approaches include (1) co-transformation with multiple transgenes comprising of separate gene constructs of two or more independent transgenes, (2) use of linked transgenes to introduce a pathway and coordinating their expression by different strategies (e.g. a plant is transformed with a single gene construct that harbours two or more linked transgenes), (3) polycistronic transgenes by expressing multiple proteins from a single promoter, (4) polyprotein expression systems and (5) chimeric transgenes for multiple gene suppression. However, 'more the better' was the trend that emerged subsequently which led to multigene engineering in plants (Naqvi et al. 2010) which was carried over and adopted till date (Curtis and Grossniklaus 2003; Vemanna et al. 2013).

10.4 Simultaneous Expression of Stress-Responsive Genes: Proof of Concept in Model Systems

Gene stacking approach by co-expressing regulatory genes would be promising for improving multiple characters simultaneously. There have been reports on the employment of stress-specific TFs in combination to combat abiotic stresses. Co-expression of *AtMYC* and *AtMYB2* in *Arabidopsis* led to enhanced expression of a few stress-responsive downstream functional genes like *rd22* and *ADH1* and resulted in enhanced osmotic stress tolerance (Abe et al. 2003). Co-expression of key detoxification pathway genes (glyoxalase I and II) resulted in enhanced salinity stress tolerance than single gene-transformed lines (Singla-Pareek et al. 2003). The transgenic tobacco plants co-expressing glyoxalase enzymes had the capacity to resist an increase in methylglyoxal, a potent cytotoxin and primary substrate for glyoxalase pathway (Yadav et al. 2005). This in turn resulted in the maintenance of higher reduced glutathione levels under salinity stress. Later, the extended suitability of this engineering strategy was demonstrated in improving heavy metal tolerance in transgenic tobacco, wherein the double transgenics were more tolerant to toxic levels of zinc when compared to single transgenic plants (Singla-Pareek et al. 2006).

Simultaneous expression of *AtbHLH7* and *AtWRKY28* in *Arabidopsis* resulted in the increased expression of genes having either of the two TF binding sites (Babitha et al. 2013). This co-expression was achieved using a modified gateway strategy which involves initial conventional cloning of individual gene cassettes into entry vectors and a one-step LR recombination reaction to create multigene expression cassette (Vemanna et al. 2013). The outcome of co-expression of *AtbHLH7* and *AtWRKY28* indicated that the expression levels of the upregulated downstream target genes was mediated either directly by these two TFs or indirectly by upregulation of other TFs. Co-expressing diverse classes of stress-responsive TFs together in crop plants to engineer multiple pathways may help in combating several abiotic stresses by regulating numerous downstream target genes. There are now evidences to indicate that the desirable levels of stress adaptation can be achieved by co-expressing a few regulatory genes (Babitha et al. 2013).

10.5 Simultaneous Expression of Stress-Responsive Genes: Case Studies in Crop Plants

There are several transgenic approaches using single candidate genes which have manipulated the cellular level tolerance through osmolyte biosynthesis (Kathuria et al. 2009), scavenging of reactive oxygen species (ROS) (Xu et al. 2013), maintenance of transcriptional machineries and cell membrane stability (Lv et al. 2009) in different plant types.

Co-expression of abiotic stress-specific TFs, *AtDREB2A*, *AtHB7* and *AtABF3* improved salinity and drought tolerance in peanut (Pruthvi et al. 2014). The transgenic peanut plants simultaneously expressing these TFs showed increased tolerance to drought, salinity and oxidative stresses compared to non-transgenic plants, with an increase in total plant biomass (Pruthvi et al. 2014). Improvement in CT traits can contribute to better performance under multiple stresses during all stages of growth and development. Interaction of multiple genes and pathways is required for overall cellular level tolerance under stress (Pruthvi et al. 2014).

In rice, co-expression of stress-responsive TFs, *EcNAC1 (NAM ATAF CUC1)*, *EcbZIP60* (basic leucine zipper 60) and *EcbHLH57* (basic helix-loop-helix67) (Babitha 2012) and regulatory gene combinations like *CcNF-YB* (nuclear factor-YB), *OsSKIP* (Ski interacting protein) and *OsSHN* (shine TF) (Nagaveni 2013) and *PgHSF4* (heat shock factor 4), *OseIF4E* (eukaryotic translation initiation factor 4E) and *Pg47* (DNA helicase 47) (Mahesh 2015), imparted tolerance to multiple stresses. Peanut transgenic plants expressing alfalfa zinc finger 1 (Alfin1), a root growth-associated transcription factor gene, *Pennisetum glaucum* heat shock factor (PgHSF4) and pea DNA helicase (PDH45) involved in protein turnover and protection showed improved tolerance and higher growth and productivity under drought stress conditions (Ramu et al. 2016). The co-expressing transgenic plants showed significant tolerance to ethrel-induced senescence and methyl viologen-induced oxidative stress (Ramu et al. 2016). Although there are studies on co-expression of stress-responsive TFs, reports on the role of basal transcriptional regulators in imparting abiotic stress tolerance are limited (Pruthvi 2013). In this direction, the combination of a basal factor, an activator and a stress-inducible TF to enhance the stability of transcriptional machinery was attempted which would regulate the relative abundance of transcripts that code for the critical proteins required for stress acclimation. The simultaneous overexpression of *AhBTF3*, *AhNF-YA7* (both basal regulators) and *EcSAP-ZF* (a stress-responsive TF) enhanced cellular tolerance to multiple abiotic stresses and simulated organellar stresses in rice genotype having drought traits (Parvathi et al. 2015). Co-expression of these genes triggered multiple determinants that indirectly helped in improving the ability of photosynthetic carbon assimilation and water accessibility in the transgenic rice. However, the exact underlying mechanisms for the superior developmental and stress-responsive phenotypes exhibited by the promising rice transgenic plants need to be unravelled by targeted *omics* analysis, since the exact targets of the basal factors are unknown.

10.6 Advantages of Multigene Stacking Strategy for Manipulation of Crop Abiotic Stress Tolerance in Crop Plants

Multiple stress tolerance can be achieved by adopting many strategies starting from conventional breeding to the most recent biotechnological interventions by developing transgenic plants using the crop genotypes having adaptive traits. A viable strategy of developing transgenic rice plants by simultaneously over expressing multiple regulatory genes in different combinations proves to be advantageous in the recent past.

As evidenced by recent findings on the emerging trend for multiple stress tolerance by over expressing one or more regulatory genes (Babitha et al. 2013; Ramu et al. 2016; Zhang et al. 2015a, b), a direct strategy of selecting transgenic lines tolerant to multiple stresses emerges to be beneficial. To assess this criterion, performing correlation analysis for the response of different transgenic lines across various stress tolerance assays is the most apt approach. However, stress tolerance coupled with higher yield/productivity will be the key towards marginal increase in productivity. The selection of the superior transgenic lines using different statistical analyses aids in the identification of those stress-specific superior transgenic lines, which are both productive and stress tolerant when compared to wild type. Devising such advantageous strategies to identify superior transgenic lines would be useful. These approaches will lead to reliable identification of elite lines which can serve as donor lines for gene pyramiding in future breeding programmes.

10.7 Targeted Trait Improvement Towards Increased Abiotic Stress Tolerance in Crop Plants Resulting from Development of Co-expression Transgenics

A highly comprehensive and orchestrated approach is needed to alter multiple traits required for combined stress tolerance in crop plants. Any approach targeted towards the mechanisms associated with cell metabolic activity would yield viable results. There are possibilities now to manipulate certain cellular level tolerance traits like protein synthesis and protein turnover, ROS detoxification, osmoregulation, etc., using transgenic approaches. In a nutshell, targeted trait manipulation can be achieved by employing co-expression of regulatory genes thereby vesting multiple stress tolerance capacity or identifying and over expressing a super regulatory gene which can govern a multitude of traits (Fig. 10.1).

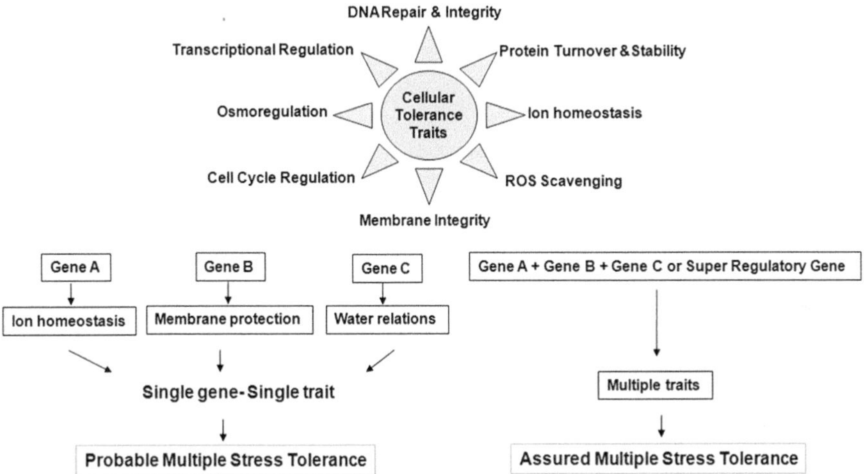

Fig. 10.1 Depiction of different cellular tolerance traits and trait manipulation for crop improvement. The different cellular tolerance traits can be manipulated by either a single gene-single trait or multiple gene-multiple trait approaches. Multiple trait manipulation results in remarkably multiple stress-tolerant genotypes, which can also be achieved by identifying a super regulatory gene which can govern the function of multiple trait-related genes

10.8 Conclusion

Simultaneous expression of trait regulatory genes relevant for multiple stress adaptation is the key towards developing elite donor lines that can be used for breeding to develop a drought-tolerant ideotype. This approach will be beneficial under multiple stress scenarios by the careful selection of crucial candidate genes for co-expression that can regulate attributing physiological processes underlying diverse stress tolerance mechanisms.

Acknowledgement This work is partly supported by the Department of Biotechnology and the Indian Council of Agricultural Research, Government of India, New Delhi. PMS would like to thank University Grants Commission, New Delhi, for providing research fellowship.

References

Abe H, Urao T, Ito T, Seki M, Shinozaki K, Yamaguchi-Shinozaki K (2003) Arabidopsis AtMYC2 (bHLH) and AtMYB2 (MYB) function as transcriptional activators in abscisic acid signaling. Plant Cell 15:63–78

Alves AAC, Setter TL (2004) Response of cassava leaf area expansion to water deficit: cell proliferation, cell expansion and delayed development. Ann Bot 94:605–613

Babitha KC (2012) Development of multiple gene construct with regulatory genes and their functional validation. Ph.D thesis, University of Agricultural Sciences, Bangalore

Babitha KC, Ramu SV, Pruthvi V, Mahesh P, Nataraja KN, Udayakumar M (2013) Co-expression of *AtbHLH17* and *AtWRKY28* confers resistance to abiotic stress in *Arabidopsis*. Transgenic Res 22:327–341

Babitha KC, Ramu SV, Nataraja KN, Sheshshayee MS, Udayakumar M (2015a) *EcbZIP60*, a basic leucine zipper transcription factor from *Eleusine coracana* L. improves abiotic stress tolerance in tobacco by activating unfolded protein response pathway. Mol Breed 35:181

Babitha KC, Vemanna RS, Nataraja KN, Udayakumar M (2015b) Overexpression of *EcbHLH57* transcription factor from *Eleusine coracana* L. in tobacco confers tolerance to salt, oxidative and drought stress. PLoS One 10(9):e0137098

Bakshi S, Dewan D (2013) Status of transgenic cereal crops: a review. Clon Transgen 3:1

Bos HJ, Tijani-Eniola H, Struik PC (2000) Morphological analysis of leaf growth of maize: responses to temperature and light intensity. NJAS – Wageningen J Life Sci 48(2):181–198

Bouaziz D, Pirrello J, Charfeddine M, Hammami A, Jbir R, Dhieb A et al (2013) Over expression of StDREB1 transcription factor increases tolerance to salt in transgenic potato plants. Mol Biotechnol 54:803–817

Budhagatapalli N, Narasimhan R, Rajaraman J, Viswanathan C, Nataraja KN (2016) Ectopic expression of AtICE1 and OsICE1 transcription factor delays stress-induced senescence and improves tolerance to abiotic stresses in tobacco. J Plant Biochem Biotechnol. doi:10.1007/s13562-015-0340-8

Cai R, Zhao Y, Wang Y, Lin Y, Peng X, Li Q et al (2014) Overexpression of a maize WRKY58 gene enhances drought and salt tolerance in transgenic rice. Plant Cell Tissue Organ Cult 119:565–577

Cai H, Tian S, Dong H, Guo C (2015) Pleiotropic effects of TaMYB3R1 on plant development and response to osmotic stress in transgenic *Arabidopsis*. Gene 558:227–234

Chen X, Wang Y, Lv B, Li J, Luo L, Lu S et al (2014) The NAC family transcription factor OsNAP confers abiotic stress response through the ABA pathway. Plant Cell Physiol 55:604–619

Comas LH, Becker SR, Cruz VMV, Byrne PF, Dierig DA (2013) Root traits contributing to plant productivity under drought. Front Plant Sci 4:442

Curtis MD, Grossniklaus U (2003) A gateway cloning vector set for high-throughput functional analysis of genes in planta. Plant Physiol 133(2):462–469

Demmig-Adams B, Adams WW III (2006) Photoprotection in an ecological context: the depleted in digalactosyldiacylglycerol. New Phytol 172:11–21

Essemine J, Govindachary S, Joly D, Ammar S, Bouzid A, Carpentier R (2012) Effect of moderate and high light on photosystem II function in *Arabidopsis thaliana*. Biochim Biophys Acta 1817:1367–1373

Feng X-M, Zhao Q, Zhao L-L, Qiao Y, Xie X-B, Li H-F et al (2012) The cold-induced basic helix-loop-helix transcription factor gene MdCIbHLH1 encodes an ICE-like protein in apple. BMC Plant Biol 12:22

Gao SQ, Chen M, Xu ZS, Zhao CP, Li L, Xu HJ et al (2011) The soybean GmbZIP1 transcription factor enhances multiple abiotic stress tolerances in transgenic plants. Plant Mol Biol 75: 537–553

Hadiarto T, Tran LP (2011) Progress studies of drought-responsive genes in rice. Plant Cell Rep 30:297–310

Halpin C (2005) Gene stacking in transgenic plants – the challenge for 21st century plant biotechnology. Plant Biotechnol J 3:141–155

Hu H, Xiong L (2014) Genetic engineering and breeding of drought-resistant crops. Annu Rev Plant Biol 65:715–741

Huang Q, Wang Y, Li B, Chang J, Chen M, Li K et al (2015) TaNAC29, a NAC transcription factor from wheat, enhances salt and drought tolerance in transgenic *Arabidopsis*. BMC Plant Biol 15:268

ISAAA (2013) Stacked traits in biotech crops. Pocket K No. 42. http://www.isaaa.org/kc

Jacobsen E, Nataraja KN (2008) Cisgenics – facilitating the second green revolution in India by improved traditional plant breeding. Curr Sci 94(11):1365–1366

Jiang Y, Yang B, Deyholos MK (2009) Functional characterization of the Arabidopsis bHLH92 TF in abiotic stress. Mol Genet Genomics 282:503–516

Joo J, Choi HJ, Lee YH, Kim YK, Song SI (2013) A transcriptional repressor of the ERF family confers drought tolerance to rice and regulates genes preferentially located on chromosome 11. Planta 238:155–170

Kamoshita A, Chandra Babu R, Boopathi NM, Fukai S (2008) Phenotypic and genotypic analysis of drought-resistance traits for development of rice cultivars adapted to rainfed environments. Field Crop Res 109:1–23

Kathuria K, Giri J, Karaba N, Nataraja KN, Murata N et al (2009) Glycine betaine-induced water stress tolerance in codA-expressing transgenic indica rice is associated with up-regulation of several stress responsive genes. Plant Biotechnol J 7:512–526

Lawlor DW (2013) Genetic engineering to improve plant performance under drought: physiological evaluation of achievements, limitations, and possibilities. J Exp Bot 64(1):83–108

Liu H, Zhou X, Dong N, Liu X, Zhang H, Zhang Z (2011) Expression of a wheat MYB gene in transgenic tobacco enhances resistance to *Ralstonia solanacearum*, and to drought and salt stresses. Funct Integr Genomics 11:431–443

Liu X, Liu S, Wu J, Zhang B, Li X, Yan Y et al (2013) Overexpression of *Arachis hypogaea* NAC3 in tobacco enhances dehydration and drought tolerance by increasing superoxide scavenging. Plant Physiol Biochem 70:354–359

Liu C, Mao B, Ou S, Wang W, Liu L, Wu Y et al (2014) OsbZIP71,a bZIP transcription factor, confers salinity and drought tolerance in rice. Plant Mol Biol 84:19–36

Liu L, Zhang Z, Dong J, Wang T (2016) Overexpression of MtWRKY76 increases both salt and drought tolerance in *Medicago truncatula*. Environ Exp Bot 123:50–58

Lu M, Ying S, Zhang DF, Shi YS, Song YC, Wang TY et al (2012) A maize stress-responsive NAC transcription factor, ZmSNAC1, confers enhanced tolerance to dehydration in transgenic *Arabidopsis*. Plant Cell Rep 31:1701–1711

Lv SL, Lian LJ, Tao PL, Li ZX, Zhang KW et al (2009) Overexpression of Thellungiella halophila H+ −PPase (TsVP) in cotton enhances drought stress resistance of plants. Planta 229:899–910

Mahesh P. (2015) Stacking of upstream regulatory genes to confer abiotic stress tolerance in rice (*Oryza sativa*). Ph.D thesis, University of Agricultural Sciences, Bangalore

Mamrutha HM, Mogili T, Jhansi Lakshmi K, Rama N, Kosma DK, Udaya Kumar M, Jenks MA, Nataraja KN (2010) Involvement of leaf cuticular wax amount and crystal morphology in regulating post harvest water loss in mulberry (Morus species). Plant Physiol Biochem 48:690–696

Meng X, Wang JR, Wang GD, Liang XQ, Li XD, Meng QW (2015) An R2R3-MYB gene, LeAN2, positively regulated the thermo-tolerance in transgenic tomato. J Plant Physiol 175:1–8

Mittler R (2006) Abiotic stress, the field environment and stress combination. Trends Plant Sci 11(1):15–19

Nagaveni BH (2013) Engineering drought tolerant traits in rice. Ph.D thesis, University of Agricultural Sciences, Bangalore

Naqvi S, Farré G, Sanahuja G, Capell T, Zhu C, Christou P (2010) When more is better: multigene engineering in plants. Trends Plant Sci 15(1):48–56

Niyogi KK, Li XP, Rosenberg V, Jung HS (2005) Is PsbS the site of non-photochemical quenching in photosynthesis? J Exp Bot 56(411):375–382

Parvathi MS, Nataraja KN (2016) Emerging tools, concepts and ideas to track the modulator genes underlying plant drought adaptive traits: an overview. Plant Signal Behav 11:1

Parvathi MS, Sreevathsa R, Rama N, Nataraja KN (2015) Simultaneous expression of *AhBTF3, AhNF-YA7* and *EcZF* modulates acclimation responses to abiotic stresses in rice (*Oryza sativa* L). Procedia Environ Sci 29:236–237

Patil M, Ramu SV, Jathish P, Rohini S, Reddy PC, Prasad TG, Udayakumar M (2014) Overexpression of AtNAC2 (ANAC092) in groundnut (*Arachis hypogaea* L.) improves abiotic stress tolerance. Plant Biotechnol Rep 8(2):161–169

Poorter H, Niinemets Ü, Poorter L, Wright IJ, Villar R (2009) Causes and consequences of variation in leaf mass per area (LMA): a meta-analysis. New Phytol 182:565–588

Pruthvi V (2013) Validation of abiotic stress responsive transcription factors by over expression in crop plants. Ph.D thesis, University of Agricultural Sciences, Bangalore

Pruthvi V, Narasimhan R, Nataraja KN (2014) Simultaneous expression of abiotic stress responsive transcription factors, *AtDREB2A, AtHB7* and *AtABF3* improves salinity and drought tolerance in peanut (*Arachis hypogaea* L.). PLoS One 9(12):e111152

Qin Y, Tian Y, Liu X (2015) A wheat salinity-induced WRKY transcription factor TaWRKY93 confers multiple abiotic stress tolerance in *Arabidopsis thaliana.* Biochem Biophys Res Commun 464:428–433

Qiu Y, Yu D (2009) Over-expression of the stress-induced OsWRKY45 enhances disease resistance and drought tolerance in *Arabidopsis*. Environ Exp Bot 65:35–47

Ramegowda V, Senthil-Kumar M, Nataraja KN, Reddy MK, Mysore KS, Udayakumar M (2012) Expression of a finger millet transcription factor, *EcNAC1*, in tobacco confers abiotic stress-tolerance. PLoS One 7(7):e40397

Rampino P, Pataleo S, Gerardi C, Mita G, Perrotta C (2006) Drought stress response in wheat: physiological and molecular analysis of resistant and sensitive genotypes. Plant Cell Environ 29:2143–2152

Ramu VS, Swetha TN, Sheela SH, Babitha CK, Rohini S, Reddy MK, Tuteja N, Reddy CP, Prasad TG, Udayakumar M (2016) Simultaneous expression of regulatory genes associated with specific drought-adaptive traits improves drought adaptation in peanut. Plant Biotechnol J 14:1008–1020

Ravikumar G, Manimaran P, Voleti SR, Subrahmanyam D, Sundaram RM, Bansal KC et al (2014) Stress-inducible expression of AtDREB1A transcription factor greatly improves drought stress tolerance in transgenic indica rice. Transgenic Res 23:421–439

Reguera M, Peleg Z, Blumwald E (2012) Targeting metabolic pathways for genetic engineering abiotic stress tolerance in crops. Biochim Biophys Acta 1819(2):186–194

Rizhsky L, Liang H, Mittler R (2002) The combined effect of drought stress and heat shock on gene expression in tobacco. Plant Physiol 130:1143–1151

Rizhsky L, Liang H, Shuman J, Shulaev V, Davletova S, Mittler R (2004) When defense pathways collide: the response of *Arabidopsis* to a combination of drought and heat stress. Plant Physiol 134:1683–1696

Rong W, Qi L, Wang A, Ye X, Du L, Liang H et al (2014) The ERF transcription factor TaERF3 promotes tolerance to salt and drought stresses in wheat. Plant Biotechnol J 12:468–479

Saad AS, Li X, Li HP, Huang T, Gao CS, Guo MW et al (2013) A rice stress-responsive NAC gene enhances tolerance of transgenic wheat to drought and salt stresses. Plant Sci 203–204:33–40

Shi W, Liu D, Hao L, Wu CA, Guo X, Li H (2014) GhWRKY39, a member of the WRKY transcription factor family in cotton, has a positive role in disease resistance and salt stress tolerance. Plant Cell Tissue Organ Cult 118:17–32

Singla-Pareek SL, Reddy MK, Sopory SK (2003) Genetic engineering of the glyoxalase pathway in tobacco leads to enhanced salinity tolerance. PNAS 100(25):14672–14677

Singla-Pareek SL, Yadav SK, Pareek A, Reddy MK, Sopory SK (2006) Transgenic tobacco overexpressing glyoxalase pathway enzymes grow and set viable seeds in zinc-spiked soils. Plant Physiol 140(2):613–623

Suzuki N, Rivero RM, Shulaev V, Blumwald E, Mittler R (2014) Abiotic and biotic stress combinations. New Phytol 203:32–43

Tirado R, Cotter J (2010) Ecological farming: drought-resistant agriculture. 02/2010; Greenpeace Research Laboratories, University of Exeter, UK, GRL-TN

Todaka D, Shinozaki K, Yamaguchi-Shinozaki K (2015) Recent advances in the dissection of drought-stress regulatory networks and strategies for development of drought-tolerant transgenic rice plants. Front Plant Sci 6:84

Vemanna RS, Babitha KC, Rao HMH, Sathyanarayanagupta SK, Sarangi SK, Nataraja KN, Udayakumar M (2013) Modified multisite gateway cloning strategy for consolidation of genes in plants. Mol Biotechnol 53(2):129–138

Vile D, Pervent M, Belluau M, Vasseur F, Bresson J, Muller B, Granier C, Simonneau T (2012) Arabidopsis growth under prolonged high temperature and water deficit: independent or interactive effects? Plant Cell Environ 35:702–718

Wahid A, Gelani S, Ashraf M, Foolad MR (2007) Heat tolerance in plants: an overview. Environ Exp Bot 61:199–223

Wasson AP, Richards RA, Chatrath R, Misra SC, Sai Prasad SV, Rebetzke GJ, Kirkegaard JA, Christopher J, Watt M (2012) Traits and selection strategies to improve root systems and water uptake in water-limited wheat crops. J Exp Bot 63(9):3485–3498

Xu J, Duan X, Yang J, Beeching JR, Zhang P (2013) Enhanced reactive oxygen species scavenging by overproduction of superoxide dismutase and catalase delays postharvest physiological deterioration of cassava storage roots. Plant Physiol 161:1517–1528

Yadav SK, Singla-Pareek SL, Reddy MK, Sopory SK (2005) Transgenic tobacco plants overexpressing glyoxalase enzymes resist an increase in methylglyoxal and maintain higher reduced glutathione levels under salinity stress. FEBS Lett 579(27):6265–6271

Zhang H, Liu W, Wan L, Li F, Dai L, Li D et al (2010) Functional analyses of ethylene response factor JERF3 with the aim of improving tolerance to drought and osmotic stress in transgenic rice. Transgenic Res 19:809–818

Zhang L, Zhang L, Xia C, Zhao G, Jia J, Kong X (2015a) The novel wheat transcription factor TaNAC47 enhances multiple abiotic stress tolerances in transgenic plants. Front Plant Sci 6:1174

Zhang L, Zhang L, Xia C, Zhao G, Liu J, Jia J et al (2015b) A novel wheat bZIP transcription factor, TabZIP60, confers multiple abiotic stress tolerances in transgenic *Arabidopsis*. Physiol Plant 153:538–554

Zhu N, Cheng S, Liu X, Du H, Dai M, Zhou DX et al (2015) The R2R3-type MYB gene OsMYB91 has a function in coordinating plant growth and salt stress tolerance in rice. Plant Sci 236:146–156

11 Tissue Water Status and Bacterial Pathogen Infection: How They Are Correlated?

Urooj Fatima and Muthappa Senthil-Kumar

Abstract

Tissue water status plays an important role in determining the fate of plant-pathogen interaction. Water availability is one of the factors that determine the multiplication of bacteria on the surface and inside the plants. Plant-water relations are highly influenced by soil water status, and drought stress is known to severely impact plant-pathogen interaction. Water, as a limiting factor, is differentially manipulated by both plants and pathogens during compatible and incompatible interactions. Plants stimulate the localized loss of water at the site of infection for limiting the bacterial multiplication. On the other hand, foliar and vascular bacterial pathogens employ different strategies to alter the plant water status and eventually establish the infection in plants. Foliar pathogens manipulate their own machinery in response to water-limited condition in plants. They also modulate the plant machinery in order to promote disease by increasing the water soaking between the cells. Similarly, vascular pathogens use different strategies such as clogging of vessels and embolism of xylem elements that leads to wilting of plant. Here, we discuss the current knowledge on impact of drought stress during plant interaction with foliar or vascular pathogen interactions.

Keywords

Apoplast • Drought • Foliar pathogen • Plant defense • Vascular pathogen • Water • Xylem

U. Fatima • M. Senthil-Kumar (✉)
National Institute of Plant Genome Research, 10531, JNU Campus,
Aruna Asaf Ali Marg, New Delhi 110 067, India
e-mail: skmuthappa@nipgr.ac.in

M. Senthil-Kumar (eds.), *Plant Tolerance to Individual and Concurrent Stresses*,
DOI 10.1007/978-81-322-3706-8_11

11.1 Introduction

Water status determines the fitness of plants and profoundly influences the outcome of plant-pathogen interaction. There are several interrelated factors that control the status of water in plant and bacterial cells. These include osmotic potential, water potential, and cell turgor pressure. Well-hydrated cell exhibits a strong turgor pressure, and any small change in water status leads to the large change in turgor pressure that may result in plasmolysis (Koch 1984; Miller et al. 1986). Plants and bacteria respond to such phenomenon by undergoing osmotic adjustment through accumulation of osmoprotectants in the cytoplasm that restore the osmotic equilibrium and maintain hydrated cell (Csonka 1989; Csonka and Hanson 1991; McNeil et al. 1999; Freeman et al. 2010; Kurz et al. 2010). Water accumulation in the cells and cell hydration are driven by the difference in the water potential of the cell and the environment. The alteration in tissue hydration is known to occur during plant-pathogen interaction. Plants and bacterial pathogens engage in struggle, in which plant restricts pathogen access to water and initiates defense response, whereas the host pathogens employ strategies to gain access to water availability and cause pathogenesis in plants. Plants employ several ways of defense against pathogens. The localized loss of water at the site of infection is the most common strategy used by plants to restrict the growth of bacterial pathogen by completely starving them of water (Beattie 2011). Mode of this manipulation involves cellular osmotic imbalance and rapid change in water potential and turgor pressure. On the other hand, bacterial pathogens need water during colonization at phyllosphere and rhizosphere and also after they gain entry inside the apoplast or vascular elements (Melotto et al. 2008). Once inside the plants, bacteria employ several strategies to acquire water. Generally, foliar pathogens in the phyllosphere are in direct contact with outside environment and deal with uneven and transient availability of water (Lindow and Brandl 2003), and these variations can have a more dramatic effect on the nutrient availability and cause desiccation stress (Leveau and Lindow 2001; Moier and Lindow 2004). The pathogens employ several strategies including the alteration of cuticle-membrane permeability (Schreiber et al. 2005) or the production of surfactants to increase the moisture on leaf surface (Bunster et al. 1989; Lindow and Brandl 2003) and the formation of bacterial aggregates (Monier and Lindow 2003). It was shown that the formation of bacterial aggregates and EPS production by *Pseudomonas syringae* pv. *syringae* B728a (causal agent of brown spot on bean plants) aid in retention of water and nutrients and increase their fitness on the leaf surface (Monier and Lindow 2003; Quiñones et al. 2005). The water availability in the apoplast could be different from the phyllosphere (Yu et al. 2013). Foliar bacterial pathogens increase the water content of the apoplast by regulating the stomatal movement and the hormonal responses. Vascular bacterial pathogens invade the xylem and disrupt the water-conducting system. This is achieved through production of toxin by bacteria or by mechanical blocking of xylem elements. This chapter covers the impact of water status on plant defense and bacterial pathogenesis. We also briefly discuss the effect of drought stress on pathogen infection in plants.

11.2 Impact of Plant Water Status on Its Interaction with Foliar Bacterial Pathogens

Leaf water status has a direct impact on bacterial multiplication in plants and plays a major role in deciding the final outcome of plant-pathogen interaction. The apoplast contains a considerable amount of water that ranges from 5 to 50% within plant and also varies with different species (Wardlaw 2005). The apoplast is the major site of colonization for foliar bacterial pathogens from species of *Pseudomonas* and *Xanthomonas* causing leaf spot disease in plants (Sattelmacher and Horst 2007; Rico and Preston 2008). Bacterial pathogen employs the strategy to manipulate the water status of apoplast by enhancing the water soaking. It has been shown that these pathogens induce water congestion in apoplast leading to severe disease symptoms associated with water-soaking lesions (Rudolph 1978). Further, it has been proved that nonpathogenic strains were able to grow in apoplast by maintaining the water congestion in the apoplast, and not only this, nutrient addition did not cause any further changes in the bacterial growth compared to water congestion alone (Young 1974). This indicates that water is the major limiting factor that determines the bacterial multiplication in plant. Bacterial pathogen employs several mechanisms for manipulating water status in the plant and alleviating water-limiting conditions. For example, several species of *P. syringae* induce the expression of genes that are involved in production of exopolysaccharide, i.e., alginate that aids in hydration by inducing water soaking in apoplast (Gross and Rudolph 1987; Keith et al. 2003). It has been shown that mutation in gene responsible for alginate production causes reduction in virulence of pathogen (Yu et al. 1999; Peñaloza-Vázquez et al. 2004). Further, a dramatic effect of water abundance in apoplast was also evident by increased virulence of *P. syringae* in *Arabidopsis* mutant carrying mutation in gene *att1* (for aberrant induction of type III genes) responsible for loose cuticle membrane and increase water permeability (Xiao et al. 2004; Tang et al. 2007). Similarly, it has been shown in another study that there is an increased multiplication for both host and nonhost bacteria belonging to *Pseudomonas* genus in *Atcyp710A1 Arabidopsis* mutants. The *AtCYP710A1* gene is involved in regulating the membrane permeability, and loss of function leads to increase membrane permeability and release of nutrients and water in apoplast from cell (Wang et al. 2012). Other studies on *Xanthomonas* showed the production of exopolysaccharide and xanthan as a strategy for water absorption that promotes the bacterial multiplication (Brunings and Gabriel 2003; Kemp et al. 2004; Lu et al. 2007; Guo et al. 2010). There are evidences that pathogens use effector molecules secreted through type III secretion system for releasing the water into the apoplast and promote their multiplication. For example, WtsE effector molecule produced by *Pantoea stewartii* (causal agent of Stewart's wilt in maize) promotes virulence by inducing intercellular water soaking (Ham et al. 2006).

High humidity and rain enhance stomatal opening for entry of pathogen (Melotto et al. 2008). Foliar pathogens employ strategies for the regulation of stomatal opening that facilitate their entry inside the plant and maintain the water status

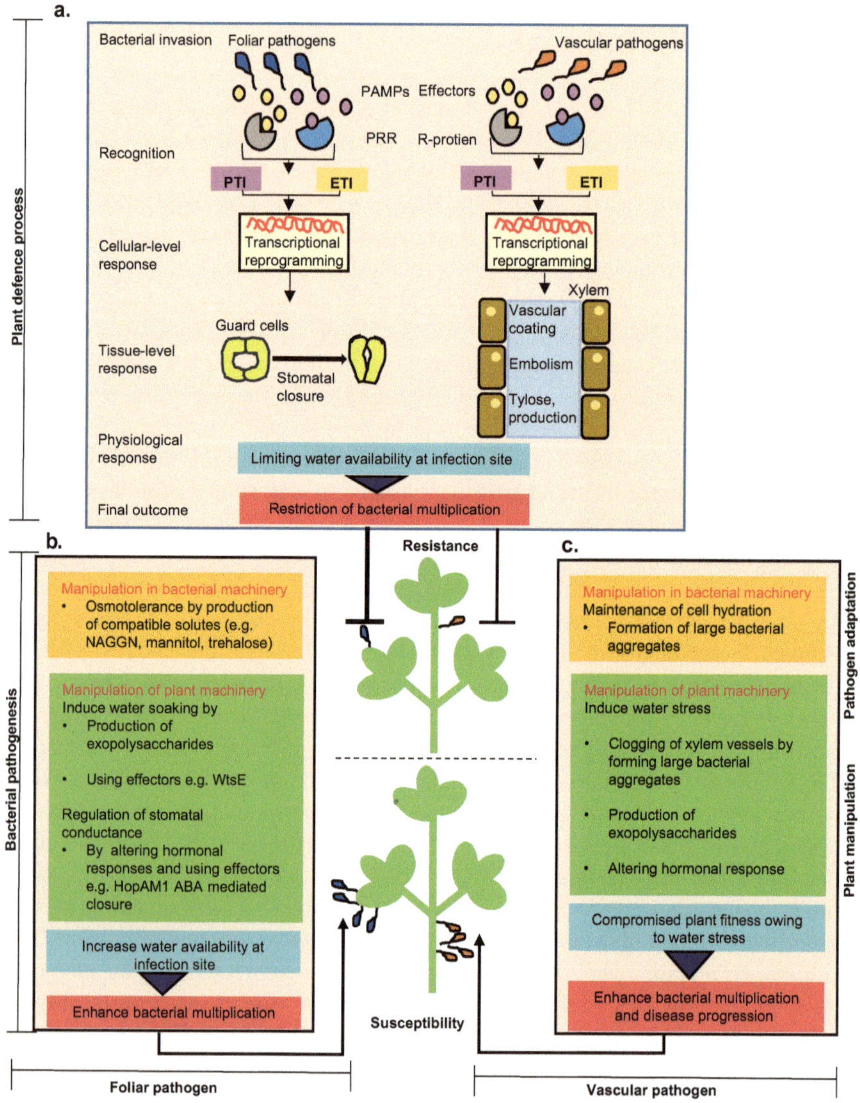

Fig. 11.1 Overview of counter strategies for regulating the water status in plant tissues for successful plant defense and bacterial pathogenesis. (**a**). Schematic representation of events occurring during plant defense response against foliar and vascular bacterial pathogens that involve limiting the water availability to pathogens at infection site, thereby restricting the multiplication of bacteria inside plants. Foliar pathogens invade the intercellular spaces inside the plant, whereas, vascular pathogens invade the xylem vessel and tracheid elements. After invasion, either the PAMP (pathogen-associated molecular patterns) molecules from pathogens are recognized by the PRR (pathogen recognition receptor) leading to PTI (PAMP-triggered immunity) or the effector molecules secreted by pathogen are recognized by R-protein leading to ETI (effector-triggered immunity). These responses lead to transcription reprogramming inside the plant cell that eventually culminates into tissue-specific response at the site of infection.

in apoplast after invasion. Bacterial pathogens secrete phytotoxins, e.g., coronatine which is secreted by *P. syringae* pv. *tomato* DC3000 (causal agent of bacterial speck of tomato) that promotes transient stomatal opening (Freeman and Beattie 2009). During early events of pathogen interaction, the induced stomatal opening helps in gaining more entry inside the plant. But later on bacteria shut the stomata to limit water loss that influences the leaf hydration. Abscisic acid (ABA) plays a main role in regulating stomatal closure in drought-stressed plants (Swamy and Smith 1999). This is supported by the evidence that due to exogenous application of ABA, the drought-stressed plants develop the susceptible response toward an avirulent pathogen, *P. syringae* pv. *tomato 1065* (Mohr and Cahill 2003). This also suggests the possibility that under drought stress condition, bacteria may have strategies to regulate the ABA-mediated stomatal closure to limit the water loss. A study showed that HopAM1, an effector from *P. syringae* pv. *tomato* DC3000, increases the multiplication and virulence of weak pathogen, *P. syringae* pv. *maculicola* M6 CE (causal agent of bacterial leaf spot on cruciferous plants) in drought-stressed plant by stimulating the ABA-mediated stomatal closure (Goel et al. 2008). This indicates that during drought conditions in plant, bacterial pathogens regulate the stomatal closure to maintain the water abundance in the apoplast for achieving successful pathogenesis (Fig. 11.1).

Further, under drought-stressed conditions, there is a cross talk between hormonal signaling pathways during pathogen infection (Pandey et al. 2014). Under drought-stressed conditions, ABA accumulates and induces stomatal closure and suppresses defense responses (Fujita et al. 2006). Studies showed that endogenous application of ABA leads to reduced salicylic acid (SA) accumulation and causes enhanced susceptibility to many pathogens (Mohr and Cahill 2003; de Torres-Zabala et al. 2007). There is evidence from few studies on *cpr22* (constitutive expressor of *PR* genes 22) and *cpn1-1* (copine1-1) lesion mimic mutants exhibit ABA-insensitive phenotypes including enhanced water loss and decrease in stomatal closure and reduced disease susceptibility when transferred from high to low humidity (Mosher et al. 2010). This suggests that ABA insensitivity leads to increased

Fig. 11.1 (continued) During foliar pathogen infection, plant defense response involves the stomatal closure to limit the water availability in the apoplast at the site of infection. During vascular pathogen infection, plant defense response involves the vascular restriction by increasing the vascular coating, embolism, and tylose production that limit the availability of water at the infection site and eventually reduce the bacterial multiplication. (**b**) This represents the foliar pathogen strategies involving its own adaptation and plant manipulation for increasing water availability at the infection site that enhance the bacterial multiplication in plants. (**c**) This represents the vascular pathogen strategies involving its own adaptation for maintaining cell hydration and manipulation of plant machinery to create water stress condition in plants. Plant diverts its energy to combat the water stress that provides opportunity for pathogen to infect more and enhance disease progression. *Left portion* indicates the responses during plant-foliar bacterial pathogen interactions, whereas, *right portion* indicates responses during plant-vascular bacterial pathogen interactions. Resistance response indicated by plant picture present *above the dotted line*. Susceptible response indicated by plant picture present *below the dotted line*

water loss due to partial closing of stomata and intense SA defense response due to loss of ABA-mediated inhibition of defense response. The cross talk between different pathways involving ABA and SA plays a major role in controlling the water status of leaf during drought stress and pathogen infection. Calcium is the common signal involved in the cross talk between different signaling pathways induced during drought stress and pathogen infection (Takahashi et al. 2011). Calcium-dependent protein kinase (CDPKs) and mitogen-activated protein kinases (MAPKs) are the potential downstream component involved in signaling pathways (Hoyos and Zhang 2000). In *Arabidopsis*, AtMAPK4 and AtMAPK6 both are activated during pathogen infection and drought stress (Ichimura et al. 2000; Desikan et al. 2001). However, the mechanisms are still unknown about how the cross talk is exploited by bacterial pathogens to enhance their virulence during drought-stressed conditions.

Osmotolerance is another important strategy used by bacterial pathogens for their survival in water-limiting condition in apoplast. Several species of *Pseudomonas* overcome the water limitation in the apoplast by inducing the expression of biosynthetic genes involved in production of compatible solutes such as N-acetyl glutaminyl glutamine amide (NAGGN), trehalose, and mannitol (Dinnibier et al. 1988; Cayley et al. 1992; D'Souza-Ault et al. 1993; Kets et al. 1996; Freeman et al. 2010). The genome analysis of *P. syringae* pv. *tomato* DC3000 and *P. syringae* pv. *syringae* strain B728a revealed the presence of large number of genes that are involved in adaptation to osmotic stress. They also have gene-encoding transporters that are involved in transport of compatible solutes and induce their expression during osmotic stress (Feil et al. 2005).

11.3 Impact on Water Status During Interaction of Vascular Bacterial Pathogens with Plants

Vascular bacterial pathogens are among the most devastating ones that specifically target xylem vessel and treachery elements that are involved in transport of water and minerals from root to different photosynthetic organs. However, very few pathogens can target living tissues because these tissues have high osmotic pressure which makes it difficult for pathogen to penetrate inside it. Since xylem is composed of relatively nutrient-poor dead plant tissues, pathogens targeting xylem are well adapted to survive in such environments and cause serious wilt disease. The pathogen causes wilt by blocking the xylem vessels and creates the water stress in plant by disrupting the water flow. Plant fitness is compromised by water stress, and plant diverts its energy to protect itself from water stress instead of pathogen infection. This provides an opportunity for pathogens to invade other tissues which are otherwise not easily accessible (Fig. 11.1). The bacterial pathogen employs several strategies for xylem blockage. The strategy involves the production of hygroscopic polymers that occlude the xylem vessels. Xylem-inhabiting pathogens such as *Ralstonia solanacearum* (causal agent of bacterial wilt disease in tomato), *Clavibacter michiganensis* subsp. *michiganensis* (causal agent of bacterial canker in tomato), and *P. stewartii* subsp. *stewartii* (Mansfield et al. 2012), secrete large amount of exopolysaccharide that disrupts the water flow and eventually blocks the xylem (Coplin and Majerczak 1990; Leigh and Coplin 1992). Among xylem-limited bacterial pathogens, *Xylella fastidiosa* (causal agent of Pierce's disease in grape and variegated chlorosis disease in citrus) is of particular importance as it has diverse and vast host range causing leaf scorching, scalding, and stunting disease (Purcell and Hopkins 1996). *X. fastidiosa* uses various strategies for xylem dysfunction such as mechanical blockage by accumulation of bacterial cells in the form of large aggregates and through embolism. Few studies showed that increase bacterial number in xylem corresponds to the water stress due to clogging of vessels. For example, alfalfa plants infected with *X. fastidiosa* induce water stress. The water stress is related with increased bacterial number in the xylem which was demonstrated by carbon isotoping in uninfected and infected plants (Daugherty et al. 2010). Similarly, in another study grapevine infected with green fluorescent protein (GFP) labeled *X. fastidiosa* reported the formation of large bacterial aggregates inside the xylem. The more occurrence of bacterial aggregates is directly related with clogging of vessels and severe water stress in xylem (Newman et al. 2003). Further, the clogging of vessel by production of polysaccharide is the most common strategy used by many pathogens. However, *X. fastidiosa* does not produce polysaccharides, but during xylem colonization, it secretes cell wall-degrading enzymes that can act on xylem pit membrane which can produce large amount of plant cell wall breakdown products in the form of polysaccharides. These can potentially clog the xylem vessels and induce the water stress (Pérez-Donoso et al. 2010). The production of tylose, the outgrowth of xylem parenchyma cells, is the most common phenomenon that occurs at the site of xylem infection. Tylose accumulation is a part of plant defense response. However, it has been shown that it disrupts the water flow in xylem and enhances the disease (Sun et al. 2013).

In addition to clogging of vessels, *X. fastidiosa* can induce embolism that disrupts the water flow in xylem of host plants such as grapevine (Newman et al. 2003; Pérez-Donoso et al. 2007). Embolism is the formation of air bubble in the vessels that disrupts the water-conducting system and induces water stress (Pérez-Donoso et al. 2010). Sometimes, embolism can be a defense response of plant against pathogen that occurs during early stages of infection, but primarily it is often a pathogen virulence strategy. Embolism may also enhance the production of tyloses that further promotes the water blockage and enhance the disease susceptibility (Stevenson et al. 2004). Further, the hormonal regulation is also involved during vascular pathogen infection and water stress. *X. fastidiosa* and *C. michiganensis* subsp. *michiganensis* infection have been associated with increased ethylene synthesis (Pérez-Donoso et al. 2007; Balaji et al. 2008). A study showed that embolism is induced in vessels of grapevine plants after infection with *X. fastidiosa* or ethylene treatment (Pérez-Donoso et al. 2007). Similarly in another study, *C. michiganensis* subsp. *michiganensis* infecting tomato plants induce the expression of ethylene responsive genes (Balaji et al. 2008). These studies indicate involvement of ethylene in inducing the drought stress and pathogenesis; however the mechanism is still unclear.

11.4 Impact of Water Status on Plant Defense Response

Plant employs several defense strategies for combating invading pathogens. Generally, plant defense response can be divided into two forms. PAMP-triggered immunity (PTI) provides resistance to wide array of pathogens, and effector-triggered immunity (ETI) is specific to a particular pathogen that secretes specific effector molecules (Jones and Dangl 2006). These defense responses operate through various means, of which limiting water availability is the most effective strategy that involves the localized desiccation at the site of infection which ultimately restricts the bacterial multiplication. Stomatal conductance is critically regulated by plant for modulating the water status in order to restrict the growth of pathogens. During PTI, plants induce the stomatal closure and affect the water availability in the apoplast at the infection site (Melotto et al. 2006, 2008). However, the alteration in water content at the site of infection is more likely determined by the rate of water depletion through stomata and water replacement through xylem. Further, it has been shown that during PTI, xylem conductivity is affected and causes a rapid decrease in vascular activity after infection. For example, the level of vascular dye accumulation was observed in minor veins after infiltration of several PAMP molecules from pathogenic and nonpathogenic strains in *Nicotiana benthamiana* plants. The reduced accumulation of dyes was seen that indicates the decrease conductivity in the xylem (Oh and Collmer 2005).

The effector-mediated plant defense response against pathogens involves the stomatal closure, loss of water from apoplast, and decrease in the water conductance in xylem. Plant induces stomatal closure after recognition of effector molecules secreted by bacterial pathogen. For example, the stomatal closure and decrease in transpiration rate were observed in *Arabidopsis* plants after syringe infiltrations of *P. syringae* strains expressing avrRpt2 and avrRpm1 indicating the impact of stomatal regulation in limiting apoplastic water availability to pathogens at infection site (Freeman and Beattie 2009). These defense responses are also modulated by the apoplastic water availability. The effector-mediated immune response triggers the hypersensitive response (HR) that causes programmed cell death at infection site. Several studies revealed that high relative humidity suppresses the HR development in plants infected with pathogens. For example, water soaking in pepper leaf infected with *X. campestris* pv. *vesicatoria* (causal agent of bacterial spot of pepper and tomato) showed delayed HR. Effector molecules secreted by pathogen in the apoplast are generally translocated to plant cell membrane. More water content in the apoplast restricted the attachment of these effectors to host membrane leading to delayed HR symptoms (Cook and Stall 1977). Similarly, high relative humidity during *Arabidopsis* infection with *P. syringae* pv. *tomato* DC3000 also delayed HR symptoms (Jambunathan et al. 2001). Hence, high water content in apoplast leads to suppression of ETI responses. This is supported by other studies using water stress-responsive biosensor which sense the water potential in apoplast of *Arabidopsis* plants after infection with *P. syringae* strains (Axtell and Beattie 2002). It was shown in susceptible interactions during early period of infection; the water potential was high, but as the ETI response is established, there was a rapid decrease in water potential indicating the influence of water level on pathogen growth in apoplast. Further, it was also shown that water potential was decreased in the apoplast after infection with nonpathogenic strains of *P. fluorescens* and type III secretion system mutant of *P. syringae* pv. *tomato* DC3000 (Freeman and Beattie 2009). It indicates that basal defense is not related with modulating the water content in apoplast; however during ETI, water availability in apoplast is limited in order to restrict the growth of bacterial pathogen (Wright and Beattie 2004). The reduction in xylem conductance can be a mechanism for limiting the water availability in the apoplast. Studies indicate that ETI leads to reduction in xylem conductance. For example, *Arabidopsis* plants infected with avirulent *P. syringae* pv. *tomato* DC3000 expressing *avrRpm1* or *avrRPt2* encounter complete loss of vascular activity indicated by vascular dye at infection site, but there was no change after infection with virulent pathogen, and the water potential in the apoplast during these interactions was very low (Wright and Beattie 2004; Freeman and Beattie 2009). This indicates that reduction in xylem conductance limits the water availability in the apoplast during pathogen infection.

11.5 Water Deficit Stress Can Be Endurance or Predisposing Factor to Pathogen Infection

Water deficit stress dramatically affects the consequences of plant-bacterial pathogen interaction. It can act as an endurance factor for plant contributing toward the resistance against pathogen or can be predisposing that enhances the susceptibility of plant toward bacterial pathogen (Bostock et al. 2014; Ramegowda and Senthil-Kumar 2015). Pathogen may undergo several physiological adaptations during water stress conditions in plant. Majority of foliar pathogens adapt several strategies for increasing their fitness on the leaf surfaces. However, the main challenge faced by pathogen on the leaf surface is to maintain proper hydration of the cells and water and nutrient supply that enhances bacterial multiplication. Under water deficit stress conditions, the pathogen compromises its physiological adaptions to survive on the leaf surface. Several studies showed that low humidity conditions on the leaf surface lead to dramatic reduction in bacterial multiplication (Beattie and Lindow 1994; Monier and Lindow 2003). Water stress condition on the leaf surface can be an endurance factor for plant to limit the bacterial multiplication and reduce the bacterial inoculum on the leaf surface. The water abundance on the leaf surface enhances the opportunities for pathogen entry inside the plant. Therefore, reduced bacterial inoculum on the surface due to water stress conditions can severely impact the amount of inoculum entering inside the plant. Plant may resist the relatively low level of bacterial inoculum, eventually providing endurance against pathogen (Bostock et al. 2014). Further, other studies indicate that drought stress acts as an endurance factor in plant against pathogen infection. For example, moderately drought-stressed *N. benthamiana* plants infected with *P. syringae* pv. *tabaci* (causal agent of wild fire disease in tobacco) show reduced bacterial multiplication and disease symptoms (Ramegowda et al. 2013). Similarly other studies done in *Arabidopsis* plants inoculated with *P. syringae* pv. *tomato* DC3000 showed that drought stress leads to reduced pathogen infection (Gupta et al. 2016). These studies reflect that water stress acts as an endurance factor to foliar pathogen multiplication on the leaf and inside the plant.

During plant-pathogen interaction under water deficit stress conditions, plant may not adjust physiologically to the same extent as that of the pathogen, and water stress may act as a predisposing factor leading to enhanced disease susceptibility. For example, previously the role of ABA has been implicated in increasing the foliar pathogen such as *P. syringae* susceptibility under water stress conditions in *Arabidopsis* plants (Mohr and Cahill 2003). Similarly, it was shown in other studies that under moderate drought-stressed condition, *N. benthamiana* plants infected with *P. syringae* pv. *tabaci* showed reduced susceptibility, but, as the drought becomes more severe, the plant showed enhanced disease susceptibility toward pathogen (Ramegowda et al. 2013). Predisposition is also demonstrated for vascular bacterial pathogens infecting plants under water stress conditions. For example, *V. vinifera* plants infected with *X. fastidiosa* during high level of drought-stressed conditions manifest enhanced disease susceptibility (Choi et al. 2013). This can be explained as the plants prioritize its response in order to minimize the effect of water stress at the cost of defense response and divert its energy to alleviate water stress. This provides an opportunity for pathogens to invade more and increases disease severity in plants. Further studies are needed to understand the underlying mechanism involved during water stress through which it can act as endurance and predisposing factor. This will help us to manipulate such mechanisms in order to develop disease-resistant plants.

11.6 Conclusions and Future Perspectives

Plants and pathogens are involved in tug-of-war with each other where both are involved in manipulating the water status of plant for their survival. Foliar and vascular bacterial pathogens act differently in employing the strategies of manipulating water status of plant for successful pathogenesis. However, the production of exopolysaccharide is the common strategy used by both in order to retain the hydration at the site of infection. Foliar bacterial pathogens regulate stomatal opening for increasing the water availability in apoplast at the infection site. On the other hand, plant also tightly regulates the stomatal conductance to prevent bacterial entry and to limit the water availability in apoplast. Changes in apoplastic water availability have profound effect on the growth of the foliar bacterial pathogen. The localized loss of water at infection site is the most common plant defense strategy. The plant defense response especially, ETI, likely involves the limiting of water content in apoplast and thereby restricting pathogen growth. High water content in apoplast directly affects the effector-mediated plant defense response. On the other hand, foliar pathogens secrete effectors such as WtsE that can induce water soaking in apoplast and HopAM1 that targets ABA signaling pathway to increase virulence in water deficit plants. Vascular pathogens disrupt plant water status by inducing ethylene levels in plants, by clogging the xylem conductance, and through embolism formation. However, embolism formation and vascular restriction can be a part of plant defense response against pathogens.

The molecular and physiological mechanisms involved in localized loss of water from leaf at infection site and restriction of vascular conductance during plant defense are important areas of research. Our knowledge regarding how pathogen alters water status for its benefit is still unclear. These are possibly repertoires of bacterial effectors that promote their growth in apoplast by enhancing water availability. It is important to identify those candidates and also elucidate the mechanism by which they modulate water status. Also, only little is known about how the drought stress influences the defense response and pathogenesis. During this interaction, cross talk between phytohormones plays an important role. It is important to elucidate these signaling pathways in order to have complete mechanistic understanding of plant defense response during drought stress. This will help us in developing crops that can effectively withstand the pathogen infection under drought-stressed conditions.

Acknowledgments Combined stress tolerance-related projects at MSk Lab are supported by National Institute of Plant Genome Research core funding and DBT-Ramalingaswami re-entry fellowship grant (BT/RLF/re-entry/23/2012) and DBT-Innovative Young Biotechnologist Award. UF acknowledges DBT-SRF (DBT/2013/NIPGR/68) fellowship.

References

Axtell CA, Beattie GA (2002) Construction and characterization of a *proU-gfp* transcriptional fusion that measures water availability in a microbial habitat. Appl Environ Microbiol 68:4604–4612

Balaji V, Mayrose M, Sherf O, Jacob-Hirsch J, Eichenlaub R et al (2008) Tomato transcriptional changes in response to *Clavibacter michiganensis* subsp. *michiganensis* reveal a role for ethylene in disease development. Plant Physiol 146:1797–1809

Beattie GA (2011) Water relations in the interaction of foliar bacterial pathogens with plants. Annu Rev Phytopathol 49:533–555

Beattie GA, Lindow SE (1994) Survival, growth, and localization of epiphytic fitness mutants of Pseudomonas syringae mutants on leaves. Appl Environ Microbiol 60:3790–3798

Bostock RM, Pye MF, Roubtsova TV (2014) Predisposition in plant disease: exploiting the nexus in abiotic and biotic stress perception and response. Annu Rev Phytopathol 52:517–549

Brunings AM, Gabriel DW (2003) *Xanthomonas citri*: breaking the surface. Mol Plant Pathol 4:141–157

Bunster L, Fokkema HJ, Schippers B (1989) Effect of surface activity of Pseudomonas spp, on leaf wettability. Appl Environ Microbiol 55:1340–1345

Cayley S, Lewis BA, Record MT (1992) Origins of the osmoprotective properties of betaine and proline in *Escherichia coli* K-12. J Bacteriol 174:1586–1595

Choi HK, Iandolino A, Goes da Silva F, Cook D (2013) Water deficit modulates the response of *Vitis vinifera* to the Pierce's disease pathogen Xylella fastidiosa. Mol Plant-Microbe Interact 26:643–657

Cook AA, Stall RE (1977) Effects of watersoaking on response to *Xanthomonas vesicatoria* in pepper leaves. Phytopathology 67:1101–1103

Coplin DL, Majerczak DR (1990) Extracellular polysaccharide genes in *Erwinia stewartii*: directed mutagenesis and complementation analysis. Mol Plant-Microbe Interact 3:286–292

Csonka LN (1989) Physiological and genetic responses of bacteria to osmotic stress. Microbiol Rev 53:121–147

Csonka LN, Hanson AD (1991) Prokaryotic osmoregulation: genetics and physiology. Annu Rev Microbiol 45:569–606

Daugherty MP, Lopes JRS, Almeida RPP (2010) Strain-specific alfalfa water stress induced by *Xylella fastidiosa*. Eur J Plant Pathol 127:333–340

de Torres-Zabala M, Truman W, Bennett MH, Lafforgue G, Mansfield JW, Egea PR, Bogre L, Grant M (2007) *Pseudomonas syringae* pv. tomato hijacks the Arabidopsis abscisic acid signalling pathway to cause disease. EMBO J 26:1434–1443

Desikan R, Hancock JT, Ichimura K, Shinozaki K, Neill SJ (2001) Harpin induces activation of the *Arabidopsis* mitogen-activated protein kinases AtMPK4 and AtMPK6. Plant Physiol 126:1579–1587

Dinnibier U, Limpinsel E, Schmid R, Bakker EP (1988) Transient accumulation of potassium glutamate and its replacement by trehalose during adaptation of growing-cells of *Escherichia coli* K-12 to elevated sodium chloride concentrations. Arch Microbiol 150:348–357

D'Souza-Ault MR, Smith LT, Smith GM (1993) Roles of N-acetylglutaminylglutamine amide and glycine betaine in adaptation of *Pseudomonas aeruginosa* to osmotic stress. Appl Environ Microbiol 59:473–478

Feil H, Feil WS, Chain P, Larimer F, DiBartolo G, Copeland A, Lykidis A, Trong S, Nolan M, Goltsman E, Thiel J, Malfatti S, Loper JE, Lapidus A, Detter JC, Land M, Richardson PM, Kyrpides NC, Ivanova N, Lindow SE (2005) Comparison of the complete genome sequences of *Pseudomonas syringae* pv. syringae B728a and pv. tomato DC3000. Proc Natl Acad Sci U S A 102:11064–11069

Freeman BC, Beattie GA (2009) Bacterial growth restriction during host resistance to *Pseudomonas syringae* is associated with leaf water loss and localized cessation of vascular activity in *Arabidopsis thaliana*. Mol Plant-Microbe Interact 22:857–867

Freeman BC, Chen C, Beattie GA (2010) Identification of the trehalose biosynthetic loci of *Pseudomonas syringae* and their contribution to fitness in the phyllosphere. Environ Microbiol 12:1486–1497

Fujita M, Fujita Y, Noutoshi Y, Takahashi F, Narusaka Y et al (2006) Crosstalk between abiotic and biotic stress responses: a current view from the points of convergence in the stress signaling networks. Curr Opin Plant Biol 9:436–442

Goel AK, Lundberg D, Torres MA, Matthews R, Akimoto-Tomiyama C, Farmer L, Dangl JL, Grant SR (2008) The *Pseudomonas syringae* type III effector HopAM1 enhances virulence on waterstressed plants. Mol Plant-Microbe Interact 21:361–370

Gross M, Rudolph K (1987) Demonstration of levan and alginate in bean plants (*Phaseolus vulgaris*) infected by *Pseudomonas syringae* pv. *phaseolicola*. J Phytopathol 120:9–19

Guo Y, Sagaram US, Kim JS, Wang N (2010) Requirement of the *galU* gene for polysaccharide production by and pathogenicity and growth in planta of *Xanthomonas citri* subsp. *citri*. Appl Environ Microbiol 76:2234–2242

Gupta A, Dixit SK, Senthil-Kumar M (2016) Drought stress predominantly endures *Arabidopsis thaliana* to pseudomonas syringae infection. Front Plant Sci 7:808

Ham JH, Majerczak DR, Arroyo-Rodriguez AS, Mackey DM, Coplin DL (2006) WtsE, an AvrEfamily effector protein from *Pantoea stewartii* subsp. *stewartii*, causes disease-associated cell death in corn and requires a chaperone protein for stability. Mol Plant-Microbe Interact 19:1092–1102

Hoyos ME, Zhang S (2000) Calcium-independent activation of salicylic acid-induced protein kinase and a 40-kilodalton protein kinase by hyperosmotic stress. Plant Physiol 122:1355–1364

Ichimura K, Mizoguchi T, Yoshida R, Yuasa T, Shinozaki K (2000) Various abiotic stresses rapidly activate *Arabidopsis* MAP kinases ATMPK4 and ATMPK6. Plant J 24:655–665

Jambunathan N, Siani JM, McNellis TW (2001) A humidity-sensitive *Arabidopsis* copine mutant exhibits precocious cell death and increased disease resistance. Plant Cell 13:2225–2240

Jones JD, Dangl JL (2006) The plant immune system. Nature 444:323–329

Keith RC, Keith LM, Hernández-Guzmán G, Uppalapati SR, Bender CL (2003) Alginate gene expression by *Pseudomonas syringae* pv. *tomato* DC3000 in host and non-host plants. Microbiology 149:1127–1138

Kemp BP, Horne J, Bryant A, Cooper RM (2004) *Xanthomonas axonopodis* pv. *manihotis gumD* gene is essential for EPS production and pathogenicity and enhances epiphytic survival on cassava (*Manihot esculenta*). Physiol Mol Plant Pathol 64:209–218

Kets EP, Galinski EA, de Wit M, de Bont JA, Heipieper HJ (1996) Mannitol, a novel bacterial compatible solute in *Pseudomonas putida* S12. J Bacteriol 178:6665–6670

Koch AL (1984) Shrinkage of growing *Escherichia coli* cells by osmotic stress. J Bacteriol 159:919–924

Kurz M, Burch AY, Seip B, Lindow SE, Gross H (2010) Genome-driven investigation of compatible solute biosynthesis pathways of *Pseudomonas syringae* pv. *syringae* and their contribution to water stress tolerance. Appl Environ Microbiol 76:5452–5462

Leigh JA, Coplin DL (1992) Exopolysaccharides in plant-bacterial interactions. Annu Rev Microbiol 46:307–346

Leveau JH, Lindow SE (2001) Appetite of an epiphyte: quantitative monitoring of bacterial sugar consumption in the phyllosphere. Proc Natl Acad Sci U S A 98:3446–3453

Lindow SE, Brandl MT (2003) Microbiology of the phyllosphere. Appl Environ Microbiol 69:1875–1883

Lu GT, Ma ZF, Hu JR, Tang DJ, He YQ et al (2007) A novel locus involved in extracellular polysaccharide production and virulence of *Xanthomonas campestris* pathovar *campestris*. J Microbiol 153:737–746

Mansfield J, Genin S, Magori S, Citovsky V, Sriariyanum M, Ronald P et al (2012) Top 10 plant pathogenic bacteria in molecular plant pathology. Mol Plant Pathol 13:614–629

McNeil SD, Nuccio ML, Hanson AD (1999) Betaines and related osmoprotectants. Targets for metabolic engineering of stress resistance. Plant Physiol 120:945–949

Melotto M, Underwood W, Koczan J, Nomura K, He SY (2006) Plant stomata function in innate immunity against bacterial invasion. Cell 126:969–980

Melotto M, Underwood W, He SY (2008) Role of stomata in plant innate immunity and foliar bacterial diseases. Annu Rev Phytopathol 46:101

Miller KJ, Kennedy EP, Reinhold VN (1986) Osmotic adaptation by gram-negative bacteria: possible role for periplasmic oligosaccharides. Science 231:48–51

Mohr PG, Cahill DM (2003) Abscisic acid influences the susceptibility of *Arabidopsis thaliana* to *Pseudomonas syringae* pv.tomato and *Peronospora parasitica*. Funct Plant Biol 30:461–469

Moier JM, Lindow SE (2004) Frequency, size, and localization of bacterial aggregates on bean leaf surfaces. Appl Environ Microbiol 70:346–355

Monier JM, Lindow SE (2003) Differential survival of solitary and aggregated bacterial cells promotes aggregate formation on leaf surfaces. Proc Natl Acad Sci U S A 100:15977–15982

Mosher S, Moeder W, Nishimura N, Jikumaru Y, Joo S-H et al (2010) The lesion-mimic mutant *cpr22* shows alterations in abscisic acid signaling and abscisic acid insensitivity in a salicylic acid–dependent manner. Plant Physiol 152:1901–1913

Newman KL, Almeida RPP, Purcell AH, Lindow SE (2003) Use of a green fluorescent strain for analysis of *Xylella fastidiosa* colonization of *Vitis vinifera*. Appl Environ Microbiol 69:7319–7327

Oh HS, Collmer A (2005) Basal resistance against bacteria in *Nicotiana benthamiana* leaves is accompanied by reduced vascular staining and suppressed by multiple *Pseudomonas syringae* type III secretion system effector proteins. Plant J 44:348–359

Pandey P, Sinha R, Mysore KS, Senthil-Kumar M (2014) Impact of concurrent drought stress and pathogen infection on plants. In: Mahalingam R (ed) Combined stresses in plants: physiological, molecular, and biochemical aspects. Springer, Cham

Peñaloza-Vázquez A, Fakhr MK, Bailey AM, Bender CL (2004) AlgR functions in *algC* expression and virulence in *Pseudomonas syringae* pv. *syringae*. J Microbiol 150:2727–2737

Pérez-Donoso AG, Greve LC, Walton JH, Shackel KA, Labavitch JM (2007) *Xylella fastidiosa* infection and ethylene exposure result in xylem and water movement disruption in grapevine shoots. Plant Physiol 143:1024–1036

Pérez-Donoso AG, Sun Q, Roper MC, Greve LC, Kirkpatrick B, Labavitch JM (2010) Cell wall–degrading enzymes enlarge the pore size of intervessel pit membranes in healthy and *Xylella fastidiosa* infected grapevines. Plant Physiol 152:1748–1759

Purcell AH, Hopkins DL (1996) Fastidious xylem-limited bacterial plant pathogens. Annu Rev Phytopathol 34:131–151

Quiñones B, Dulla G, Lindow SE (2005) Quorum sensing regulates exopolysaccharide production, motility, and virulence in Pseudomonas syringae. Mol Plant-Microbe Interact 18:682–693

Ramegowda V, Senthil-Kumar M (2015) The interactive effects of simultaneous biotic and abiotic stresses on plants: mechanistic understanding from drought and pathogen combination. J Plant Physiol 176:47–54

Ramegowda V, Senthil-Kumar M, Ishiga Y, Kaundal A, Udayakumar M, Mysore KS (2013) Drought stress acclimation imparts tolerance to Sclerotinia sclerotiorum and Pseudomonas syringae in Nicotiana benthamiana. Int J Mol Sci 14:9497–9513

Rico A, Preston GM (2008) *Pseudomonas syringae* pv. tomato DC3000 uses constitutive and apoplast-induced nutrient assimilation pathways to catabolize nutrients that are abundant in the tomato apoplast. Mol Plant-Microbe Interact 21:269–282

Rudolph K (1978) Host specific principle from Pseudomonas-phaseolicola (Burkh) Dowson, inducing water-soaking in bean-leaves. Phytopathologische Zeitschrift- J Phytopathol 93:218–226

Sattelmacher B, Horst WJ (2007) The apoplast of higher plants: compartment of storage. In: Sattelmacher B, Horst WJ (eds) Transport and reactions: the significance of the apoplast for the mineral nutrition of higher plants. Springer, Berlin

Schreiber L, Krimm U, Knoll D, Sayed M, Auling G, Kroppenstedt RM (2005) Plant microbe interactions: identification of epiphytic bacteria and their ability to alter leaf surface permeability. New Phytol 166:589–594

Stevenson JF, Matthews MA, Greve LC, Labavitch JM, Rost TL (2004) Grapevine susceptibility to Pierce's disease II: progression of anatomical symptoms. Am J Enol Vitic 55:238–245

Sun Q, Sun Y, Walker MA, Labavitch JM (2013) Vascular occlusions in grapevines with Pierce's disease make disease symptom development worse. Plant Physiol 161:1529–1541

Swamy PM, Smith B (1999) Role of abscisic acid in plant stress tolerance. Curr Sci 76:1220–1227

Takahashi F, Mizoguchi T, Yoshida R, Ichimura K, Shinozaki K (2011) Calmodulin-dependent activation of MAP kinase for ROS homeostasis in Arabidopsis. Mol Cell 41:649–660

Tang D, Simonich MT, Innes RW (2007) Mutations in *LACS2*, a long-chain acyl-coenzyme a synthetase, enhance susceptibility to avirulent *Pseudomonas syringae* but confer resistance to *Botrytis cinerea* in *Arabidopsis*. Plant Physiol 144:1093–1103

Wang K, Senthil-Kumar M, Ryu CM, Kang L, Mysore KS (2012) Phytosterols play a key role in plant innate immunity against bacterial pathogens by regulating nutrient efflux into the apoplast. Plant Physiol 158:1789–1802

Wardlaw IF (2005) Consideration of apoplastic water in plant organs: a reminder. Funct Plant Biol 32:561–569

Wright CA, Beattie GA (2004) *Pseudomonas syringae* pv. *tomato* cells encounter inhibitory levels of water stress during the hypersensitive response of *Arabidopsis thaliana*. Proc Natl Acad Sci U S A 101:3269–3274

Xiao F, Goodwin SM, Xiao Y, Sun Z, Baker D (2004) *Arabidopsis CYP86A2* represses *Pseudomonas syringae* type III genes and is required for cuticle development. EMBO J 23:2903–2913

Young JM (1974) Effect of water on bacterial multiplication in plant tissue. N Z J Agric Res 17:115–119

Yu J, Peñaloza-Vázquez A, Chakrabarty AM, Bender CL (1999) Involvement of the exopolysaccharide alginate in the virulence and epiphytic fitness of *Pseudomonas syringae* pv. *syringae*. Mol Microbiol 33:712–720

Yu X, Lund SP, Scott RA, Greenwald JW, Records AH, Nettleton D, Lindow SE, Grossd DC, Beattiea GA (2013) Transcriptional responses of *Pseudomonas syringae* to growth in epiphytic versus apoplastic leaf sites. Proc Natl Acad Sci U S A 110:e425–e434